PHILOPONUS

On Aristotle Physics 4.6-9

PHILOPONUS

On Aristotle Physics 4.6-9

Translated by
Pamela Huby

B L O O M S B U R Y
LONDON • NEW DELHI • NEW YORK • SYDNEY

Bloomsbury Academic
An imprint of Bloomsbury Publishing Plc

50 Bedford Square
London
WC1B 3DP
UK

1385 Broadway
New York
NY 10018
USA

www.bloomsbury.com

Bloomsbury is a registered trade mark of Bloomsbury Publishing Plc

First published in 2012 by Bristol Classical Press
Paperback edition first published 2014

British Library Cataloguing-in-Publication Data
A catalogue record for this book is available from the British Library.

ISBN HB: 978-1-7809-3091-6
PB: 978-1-4725-3916-8
ePDF: 978-1-4725-0176-9

Library of Congress Cataloging-in-Publication Data
A catalog record for this book is available from the Library of Congress.

The present translations have been made possible by generous and imaginative funding from the following sources: the National Endowment for the Humanities, Division of Research Programs, an independent federal agency of the USA; the Leverhulme Trust; the British Academy; the Jowett Copyright Trustees; the Royal Society (UK); Centro Internazionale A. Beltrame di Storia dello Spazio e del Tempo (Padua); Mario Mignucci; Liverpool University; the Leventis Foundation; the Arts and Humanities Research Council; Gresham College; the Esmée Fairbairn Charitable Trust; the Henry Brown Trust; Mr and Mrs N. Egon; the Netherlands Organisation for Scientific Research (NWO/GW); the Ashdown Trust; Dr Victoria Solomonides, the Cultural Attaché of the Greek Embassy in London. The editor wishes to thank William Charlton, Myrto Hatzimichali, Ian Crystal and Peter Lautner for their comments, Sebastian Gertz for preparing the volume for press, and Deborah Blake at Bloomsbury, who has been the publisher responsible for every volume since the first.

Typeset by Ray Davies
Printed and bound in Great Britain

Contents

Conventions

[...] Square brackets enclose words or phrases that have been added to the translation or the lemmata for purposes of clarity, as well as those portions of the lemmata which are not quoted by Philoponus.

(...) Round brackets, besides being used for ordinary parentheses, contain transliterated Greek words and Bekker page references to the Aristotelian text.

Introduction

Philoponus is one of the leading commentators on Aristotle. Born about 490, he was a student of Ammonius in Alexandria, and produced a series of lengthy commentaries aimed at students of Aristotle's works. He was a Christian, though rather an unorthodox one, and was an object of hostility to his contemporary Simplicius, a pagan Neoplatonist based in Athens. Philoponus was also a Neoplatonist, and his religion had little effect on his treatment of this part of Aristotle's *Physics,* and he comes across as an independent thinker.

In this section Aristotle is concerned with refuting the arguments for the existence of a void: the origin of these ideas is uncertain, and Aristotle is mainly concerned on the one hand with the attacks on them by Anaxagoras and Empedocles, and on the other with the views of the Atomists, who supposed that atoms moved through a void. He also deals shortly with a nebulous theory of the Pythagoreans. Philoponus' commentary on this is fairly unremarkable with the exception of two areas in which he differs from Aristotle, and one of which has been hailed as an innovation of great significance.

If we look at him as a writer of books, Philoponus is not well organised, and appears to be repetitious, but he should be seen rather as a lecturer: his method, following standard custom, is to take each part of Aristotle, as given in a lemma, and treat that as his starting-point. In fact he often gets ahead of his lemma, summarising what is to come, and he sometimes puts arguments into the logical form that would be familiar to his students, but was not Aristotle's. It is reasonable to see him as lecturing to students in a relaxed way and allowing them to ask questions and raise difficulties, to which he makes suitable replies, thus giving explanations, as of the way a clepsydra works, or of what individual words mean and why Aristotle uses them, and diverging occasionally to even more philological matters, as when he discusses the word *ekpurênizein*, which he derives from the word for an olive-stone, and the use in one place of 'but' rather than 'and'. Such remarks could be seen as replies to questions from students, and his use of puns, and, perhaps, his use of examples like that of the effects of gas in the stomach, could be seen as comic relief. Even his criticism of Aristotle in which he says that one must not be overawed by his reputation (651,3-4), might remind them of Aristotle's famous remark that Plato is dear to him, but truth is dearer, and his report of a claim of the spuriousness of a part of the text by earlier commentators would have helped his students.

The commentary is interrupted at 675,12-695,8 by what has become known as the Corollary on Void. This has been translated by David Furley and published separately in this series, and contains many of Philoponus' own thoughts. There are, however, two discussions in the main work of great importance, first a departure from Aristotle's account of motion in terms of natural place, for which Philoponus substitutes the idea that the arrangement of things in space is due to the fact that it is good for them to be so, and secondly what has been described as a theory of impetus. This is introduced as an alternative to Aristotle's eccentric theory which involves the continuation of motion of a thing thrown being dependent on the movement of pockets of air that surround it. Instead Philoponus argues that an impetus is imparted by the thrower which continues until it is exhausted, a view that survived to be taken up by Galileo, and has been seen as a Kuhnian revolution in science.

Textual Emendations

V indicates that the change was suggested by Vitelli.

613,26	read *eipon* for *eipe*
617,11	read *topon diastêma en hôi mêden esti bareos kai kouphou*
625,12	read *autôi* for *autois* V
625,21	read *eplêrou* for *eplêroun* V
633,12	repunctuated
639,3	*deiknusi* for *deiknus*
649,22	text added *hon ekhei to meros tou aeros*
657,21-2	*pros* is repeated in the text. Delete one.
664,23	interrogation mark added
666,22	remove comma
673,22	*hon*, not *on* (typo)
695,27	add *ou*

PHILOPONUS
On Aristotle
Physics 4.6-9

Translation

John Philoponus, Commentary on Aristotle's *Physics* Book 4, Chapters 6 to 9

Chapter 6

213a12 In the same way it must be thought that it is the business of the natural philosopher to enquire also about the void, whether it exists or not. 606,25

After the discussion about place he then next goes through that on void[1] because of the relationship between the subjects. For the void 607,1 and place appear to be the same to those who utterly believe that the void exists; for they say that the void is nothing other than place deprived of body.[2] It is with good reason therefore that after the account of place he at once goes through that about void. In the same way, he says, as we talked about place, in that way we will also discuss the void; for he will also enquire about void, first if it exists or not, and will argue on both sides, then he will set out the commonly held opinions about the void. But before completely putting into the discussion about it and setting out the arguments that introduce the void, he refutes those of the older thinkers that are trying in an unsatisfactory way to show that there is no void, so that, having supported as far as possible the argument which introduced the void he can then show that in no way does the void exist, so that he may not appear to condemn something without an argument. Anaxagoras[3] therefore, he says, and those who tried to show that the void did not exist did not meet the argument head on; for it was not what men take to be void that he showed not to exist, nor did he oppose the nature of the void itself, but what erratic people say the void is, that he demolished; but he should have made his objections to their idea,[4] not against the erratic supposition. For those who suppose that the void is an extension[5] believe that it is empty of all natural body, believing that all that there is is a natural body that is perceptible, and that air is not a perceptible body (for it falls under neither sight nor hearing, nor the other senses; for it does not have

perceptible differences that are obvious) and they supposed that the air is nothing at all, and for this reason they say that what is full of air is void. They ought therefore, if they were to meet this view correctly, to show that what they imagine to be void does not exist, I mean an extension deprived of all natural body; they do not prove this, but they do prove that air is some body. For by filling the wineskins, he says, with breath, they show that before they are filled with breath they are twisted and whirled around,[6] but not so after they have been filled, which would not have come about unless the air were a substantial body; for if the skin had been empty,[7] what would have prevented its being whirled round as it had before it was blown up? And again they demonstrate the same thing from the
608,1 clepsydras by taking the air into them. A clepsydra[8] is a vessel with holes diametrically opposed in which by stopping the one hole and letting it [the vessel] down into water they demonstrate that water does not enter, since it is filled inside with air, and it does not enter because two bodies cannot be in the same [place], and when they have unstopped[9] the hole, then the water enters through the other one, with the air giving way to the water by the remaining one. But to teach that the air is something is not to do away with the nature of the void; for it is possible for the void to have been interspersed among the air, as the followers of Democritus[10] said, or actually outside the heaven.[11]

'These people', he says, as I stated,[12] 'do not meet the problem head on', but those who have introduced the void appear to speak with greater reason. They try to show that the void exists through five arguments. The first is from the motion in place[13] of bodies: motion in place is twofold ([for] things either move as a whole or in parts, as wholes like those being carried from down to up or the opposite or sideways, in parts as with things that increase), now for the moment he produces an argument that there is a void from the things that move as a whole. He puts together the argument in the [form of] the first of the hypotheticals[14] like this: if there is motion and a body does not pass through a body, there is a void; but there is motion and a body does not pass through a body; there is therefore a void. For if motion exists, there is every necessity for moving things to pass either through a void or through a body; if then it is through a body, there will be two bodies in the same [place], and that is impossible;

for like that, the biggest thing will be given room in the smallest; it is left therefore that movement comes about through a void. The void therefore exists. He brings in Melissus[15] as a witness to this argument; for Melissus thought that the all was motionless for this reason, that he supposed that the void did not exist, as if motion could not come about otherwise than through a void. Melissus therefore produces an argument for obscure things by the demolition of things 609,1
that are obvious; for it is clear that motion exists and obvious to one who has sensation; the existence or not of the void is however obscure. So taking the obscure as agreed, in this way he does away with what is evident; for if there is motion, he says, there is a void, but there is no void, so there is also no motion. But with greater reason those introducing the void argue for the unknown by positing things that are agreed: for if motion exists, there is a void, but the first is so, the second therefore also. The cause of the error they made is very clear; for they did not understand the interchange of places (*antiperistasis*) by bodies. The second argument is from things that move in their parts, like things that are increasing; for there would be no increase, he says, if there were no void. For increase comes about through body with [the nourishment] being assimilated entirely to the increasing body; how then is it assimilated? For it is necessary either for the nourishment, by passing into the internal voids in the body, to be assimilated to the body in this way, or for a body to pass through a body. So that if the second is false, it is necessary for the first to be true, I mean that there is a void in the body. Against these Aristotle neatly said in the *On Coming to Be* that if increase came about like this, if the body increases in itself as a whole it will follow[16] that the whole body is void (at least if it is necessary for the nourishment to be assimilated everywhere, and it is assimilated through the void), and in addition it is not increase, but only a filling up of the voids. It is in this way that they try to demonstrate from motion in place, both as wholes and in parts, that void exists. The third is from the condensation and rarefaction of the bodies; for how, they say, would condensation have come about unless the body is compressed into the voids in it? How would rarefaction, if the parts were not separating and leaving empty spaces in between them? Clear evidence is what happens with the

wine. For the same jar, he says, will accommodate the same amount of wine that it had received in itself, [if it were] put into a skin, along with the skin, which would not have happened, they say, unless there were some voids within the wine into which the wine was compressed and contracted into a smaller volume. This only happens with the new wine, not with all wine, from which the cause is obvious. We will
610,1 talk more widely in what comes next, when he himself solves the problem. The fourth argument is from the example that comes from the water on the ashes.[17] For if you fill a cup with water, and then transfer the water into another vessel and fill the cup with ashes, then pour the water into it, the cup will take in all of the water which had filled it even before it took in the ashes, which would not have happened if there had not been some voids existing among these bodies, those of the ashes and the water. The fifth argument is from well-known persons, I mean the Pythagoreans.[18] For they say that the void is outside the universe, and is boundless, and it is breathed in by the things here and enters the heaven and makes the distinctions among things. For nothing else is a cause of these distinctions than the void; for if this had not been what prevented them, by creeping among them, from colliding with one another and being united, the all would have been one and continuous. If therefore things are distinguished from one another, and the distinction comes from the intertwining of the void [in things], the void therefore exists. Not only did they say that this universe had a share of the void, which came in from outside, but much earlier the numbers [had a share] as well (the void is a cause of the distinguishing of them also), or rather there would not be numbers at all if there were no void; for the cause of the distinguishing of the monads is the void. But it is clear that just as the Pythagoreans used to say everything by means of symbols, so they said this also in a symbolic way.[19] For they indicated the power of distinguishing what exists by the void; for it is clearly in a way a power of putting together in the all, and so of distinguishing them.

213a13 And how it is and what it is, [as it was also about place.]

How it is, whether there is some void itself in itself outside the universe and boundless, or [one] scattered among the bodies, as those who believed in the atoms said (for being kept apart within the void the atoms never coalesce or touch; so that the bodies are constituted from the atoms and the void which is spread out inside in little bits), and whether it is separate from bodies or inseparable. The 'what it is', whether it is some extension having in itself none of the natural bodies, or is matter for bodies or something else.

> **213a14** For there is about the same belief and disbelief because 611,1
> of our assumptions.

Just as, he says, place seems alike to exist and not to exist from the assumptions about it (for the arguments on either side seem to be in a way of equal strength, there being no reason to believe more that it exists or does not exist), thus also with the void the belief in its existence and the disbelief are equal from the arguments on either side. But one must not take 'about the same' precisely in the sense that the arguments on either side for place and for the void are precisely similar in their persuasive power, but that they have a close similarity with regard to the belief and disbelief in them: for the arguments in favour of the void are weaker than those that are for doing away with the existence of place.

> **213a15** For those who say that the void exists suppose that it is like a kind of place and a container.

Since he said that it is for the natural philosopher to deal with the void in the same way as with place, he backed this claim up in these [words], that those who suppose that the void exists suppose that it is like place and a container; so that the questions raised about place will also be raised about void. Then one must add 'for belief and disbelief are about the same'; for in all these ways he indicates the relationship of the enquiries.

> **213a16** It seems to be full when it includes the solid body which it can receive, and when it lacks it, it is void [as if void and full and place are the same, but their existence is not the same. The

enquiry must begin by taking what is said by those who say that it is and again what those say who say that it is not ...]

That the void and place and extension and the full are basically the same, but different in their relations and their definition. It is called extension when it is looked at in itself and apart and not in a relation, but place when it takes on a relationship to the body, and void is[20]
612,1 place devoid of body, and full being the same as place, and perhaps there is no difference, but if there must be some differentiation between them, there would be a difference in that in place the receptacle is seen as one thing and what is received as another, but in the full the whole is seen as one thing and extension in place as like matter of the body.

213a21 And third the common opinions about them. [Some try to show that they do not exist, and they refute not what men want to call void, but what those who are mistaken say, like Anaxagoras and those who refute them in this way. For they demonstrate that air is something, by making wineskins taut and showing that air is resistant.]

For even if the void does not exist and does not have any nature among things that are, yet since the mind has very imaginative ideas about it, one ought to say whatever it is that those who say that it is among the things that are suppose that it is, and what characteristics it has or does not, such as that it appears to be an extension and a place devoid of body and like a vessel with nothing in it. For someone could also ask about a goatstag,[21] although it does not exist in reality, what the imagination conceives about it.

213a27 ... and taking [it] in the clepsydras.

Clepsydras,[22] he says, or the things from ingenious people by which they produce the pipes and the other kinds of sounds, or, what is more, he mentions clepsydras, as I have already said, the vessel which many people call the snatcher;[23] for this has the mouth which is narrow, but what is at the bottom is pierced with very small holes. As to 'taking in', or taking in breath through stopping the mouth; for when, if the vessel is empty of water, we stop the mouth and let it [the vessel] down into water, it is not filled with water when the breath has been taken in, for the said reason (for either it takes in nothing at all, or very little), and when it is full of water, by taking in the water they show in the same way that the air exists; for if it did not, whyever by stopping the entry of the air do we block the route of the water, but letting go we give a passage to the water? For this comes about through nothing other than that it is not possible for a void to exist. When therefore we give a passage to the air, as filling up the place of the water that is going out, then the water goes out. So that he demonstrates in both ways that the air is something. It is more reasonable to understand 'taking in' of air than of water; for they wanted to show not that there is some extension that is void, but that the air is something (for it is for this reason that Aristotle accuses them), and through what came before it is rather shown that the air is something.

> **213a27** But men want the void to be an extension in which there is no perceptible body. [But thinking that all being is body, they say that the thing in which there is nothing at all, that is void, because what is full of air is void. One ought to show this, that the air is something ...]

Since, he says, they believe that void is what is deprived of perceptible body (he said perceptible because of the mathematical [body], and such is also the void extended three ways and without quality) since therefore they say that void is what is deprived of perceptible body, and they say that all that exists is body (for what is not body is nothing) for this reason they say that void is that in which there is nothing at all. Since therefore the air does not have obvious perceptible qualities, because of that they say that it is not anything, but that void is the extension of the air.

> **213a31** But that there is no extension other than bodies, neither separate nor actually existent, which takes apart the whole body, so that it is not continuous [as say Democritus, Leucippus and many others of the naturalists,[24] or if there is something

outside the whole body which is continuous. These men therefore did not meet the problem head on]

'Separate' instead of capable of being separated, I mean an extension that contains a body, but is capable of being separated and being void in itself; 'actually': an extension void of body. And that either scattered among the bodies and stopping them from being continuous, as the followers of Democritus and Leucippus said, or not being scattered among the bodies, but being continuous, and being a void in itself outside the heaven as above all the imagination of many has it in considering that there is some boundless void outside the heaven, and the Pythagoreans spoke like this as I have[25] already said. They say that those around Zeno of Citium[26] thought like this also.

614,1 **213b3** But those who say that it exists more.

It is more persuasively, he says, that those speak who say that the void exists, than those who oppose this view, of whom Anaxagoras was one.

213b4 They say first that movement in place would not exist [(that is motion and increase); for motion would not appear to exist, if there were no void; for the full is unable to receive anything. If it did so receive and there were two bodies in the same [place], it would be possible for any number of bodies to be together.]

Having spoken of the arguments that seem to do away with the nature of the void, and criticised those who do not make their opposition head on, he sets up now those who argue that the void exists. First the [argument] from motion in place. This is twofold, the one as wholes, the other in parts, and they get their credibility from both kinds of motion, and there are two arguments from motion. First he argues from motion as wholes.

213b8 For one cannot say what the difference would be by which what has been said would not be the case. [If it were possible, the very smallest will receive the largest; for many smalls make a large; so that if many of the same size can exist in the same [place], so will many of unequal sizes.]

Since he said that if two bodies were in the same [place], it is possible that any other number could be, he brought in the proposals by arguing in this way. It is absurd, he says, to say on the one hand that it is possible for two bodies to be in the same [place], but on the other that it is not possible for more or a limitless number; for what distinction will they provide here, between its being possible for two, but not more? For it is not fuller now when it has been doubled than it was when it was single before; for it was equally continuous and equally filled even before this, but even so it received something else too. Why then not another again and again another? And so the largest will be in the smallest; for it is possible to cut up the largest into many equal small ones. So that, he says, it is possible for many equals to be in the same [place]. Not only those, but also unequal, unequal clearly not directly (for those who say that body passes through body would not agree to this), but that it is possible for the equal to be cut up into unequals.

213b12 Melissus then proves that the all is motionless from these; [for if there is going to be motion, it is necessary (he says) for void to exist, but the void is not among the things that exist. In one way then they show that there is void, and in another it seems that some things contract and are compressed, as they say that jars take in the wine along with the wineskins, as if the body were being condensed into the voids that are in it.]

That if there is motion it is necessary for void to exist, he argues also from the theory of Melissus. He [Melissus] at any rate gives in to the argument that does away directly with the void and simultaneously does away with motion, that is, having wanted the all to be motionless he does away with the void along with motion, since along with the idea of motion there immediately comes that of the void, and conversely with the demolition of the void motion is also done away with. But Melissus did away with motion, which obviously exists, through accepting what is obscure, I mean that there is no void; it is then much more reasonable for people to yield to what is obvious and deduce the obscure from it.

> **213b18** And again increase can[27] come about through a void; [for nourishment is body, and two bodies cannot exist together. They give as evidence also the case of the ashes, which take in the same amount of water as the empty vessel. The Pythagoreans also said that there was a void.]

The third argument, that from increase, and that is increase and motion in parts. For if the increase comes about with the nourishment being attached to what is increasing, and the nourishment is a body, and the body increases as a whole through a whole, and it is impossible for the nourishment to be joined to the body as a whole (for there will be a body in a body), it is therefore necessary for there to be some voids in a body, into which the nourishment passes.

> **213b23** And that it came into the very heaven from the boundless breath, as if it were breathing in also the void, which separates off the natures.

Supposing equally that the boundless outside the heaven was something void, this very thing they also called breath, as if both void and breath were spoken of side by side. He says therefore that this void comes in from what is outside the heaven into the heaven, as if the heaven breathes in the void, and that the heaven breathes out to
616,1 distinguish the natures, as if, he says, the void is a cause of the distinguishing of things.

> **213b25** As if the void was a kind of separation between successive things and a division. [And this is primary among numbers; for the void determines their nature. The reasons from which some say [the void] exists and some say [that it does] not are of this kind and so many.]

In the fifth book [226b34-227a1] he defines 'successive' as of things where there is nothing of the same kind in between. The void, therefore, he says, is the cause of successive items being continuous, but separated. For the void is a separation of them and a distinction, and a cause of their difference.

Chapter 7

> **213b30** For deciding whether it exists we need to bring in the meaning of its name. [The void seems to be a place in which there is nothing.]

He himself in his *Apodeictic*[28] said that when we are discussing the meaning of a word with several senses, we ought first to distinguish the different meanings of the word, and then to decide which of them our argument is about, and following after that it remains to proceed according to the problems, first enquiring if there is such a thing as is indicated by this word, then what it is, of what kind it is, what is its source. Thus, if the enquiry is about a goat-stag we must first ask whatever is meant by this word, and if there are several meanings, which of them our present enquiry is about, and after finding that out we will ask next if there is such a thing as is meant by the word, and then the rest accordingly.

Thus then here too Aristotle sets out the arguments on both sides, both of those who appear to be arguing that the void exists, and of those who are denying it, and, turning to setting out the true opinion, asks what the name of the void means. For it does not seem to be simple. For it is also related to matter, as he himself will go on to say. Some, he says, state that the void is that in which nothing exists. For they suppose, he says, that everything that exists is body. So that 617,1
they say that that in which there is no body, is void. And they say that every body is tangible, and everything that is tangible has weight or lightness. Hence, he says, they are committed to saying, with a logical argument, that the void is that in which there is nothing heavy or light. But if the void is this in which there is nothing heavy or light, since in the point too there is nothing heavy or light, the point too would be void. But it is absurd[29] to say[30] that the point is a void, and they have been involved in this absurdity through not

adding that we say that a void is an extension in which there is nothing heavy or light; for even if they add place, saying that a void is a place in which there is nothing heavy or light, they will achieve nothing as long as they do not make a distinction, saying that it is an extension in which there is nothing of weight or lightness.[31] But even if, he says, they define the void in this way, that it is an extension in which there is nothing heavy or light, or an extension receptive of weight and lightness, they will on the one hand avoid saying that the point is a void, but on the other they will fall into an even greater trap. For if this extension, he says, does not have weight or lightness, but it does have colour or sound, that is, it is a body that is coloured or sounding,[32] but [having] neither weight nor lightness, as the heaven is, what will they say? Is a thing like this void or not? Then, speaking on their behalf, he says that perhaps they would say that if this is void, what is receptive of weight and lightness, even if it were full of some other body, they would say it was void, but if it were receptive of any body whatsoever while not yet having any body, [they would say] what was like this was not void but full. But whether they said that such a thing [were full] or void, it is absurd: for it will follow that the same thing is both void and full. For it will be void if void is an extension having nothing heavy or light, but again it will be full, because the extension is filled and cannot receive another body: for there would then be two bodies in the same [place]. Therefore the same thing would be both full and void. [Or] they would 618,1 do away with the definition they had provided when they said that it is an extension in which there is nothing either heavy or light. In one way, therefore, he says, they define the void like this, as an extension in which there is no tangible body, and tangible is what has weight or lightness, but others, he says, define void in another way, so that they may avoid these absurdities, by saying that void is an extension in which there is absolutely no body nor any bodily nature.[33] And these same people, he says, also say that the void is matter (for the latter is no actual body), and these same people also say that matter is place.[34] But to say that void is matter or place is quite absurd. 'For', he says, 'matter cannot be separated from things,[35] but void', he says, 'as separated from things' it is like this that they speak of it who say that it actually exists in itself. We say the same things also about place, and that place and the void are nothing to do with the thing in it, but matter is an element of the thing, and what else has been stated in what has gone before. And that the void is not an extension in which there is no body at all, as has just been said, he shows from that, and he uses for the demolition of the void the proofs he has given before about place.[36] For since, he says, those who say that void exists say that it is nothing other than place with extension and bereft of body, and it has been shown in the account of place that it is impossible for place to be like this, an extension other than and separate from bodies, therefore the void does not exist. For those who think in this way of place distinguish void and place from one another in thought only, as we have said, in that we think of place when it has actually received a body (for it is the place of something), but of the void when it has not yet received a body, since it belongs to both to be extensions which are different from bodies. For from the same [sources] from which came the idea that place is an extension, there also came the idea of the void. For motion in place led both to the idea of place (for when people saw moving things occupying a different place at a different time, and in the same place different bodies coming to be at different times, they came upon the idea of there being an extension which in itself received different bodies at different times) and from the same idea again they suppose that the void exists. For since there is motion, they say, and body does not pass through body, there must be a void through which what is moving will pass. [They said this] because they did not yet understand the interchange [of place] of bodies. Since then, he says, both the void and place which has extension(s) are the same in their substrate,[37] and the same cause has led us to the thought of them, with the arguments, he says, with which we refuted the view that place is extension, with these same ones we will also show that the void does not exist. For if there were a void of this kind, whenever a body comes to be in it the extensions will pass through one another, and they will divide one another to infinity,[38] and all the other points which he assembled in the section about place. We too then, with the arguments with which we refuted the arguments destroying[39] [the view that] place is extension, with these we also refuted those that destroyed [the view that] void is as it is said to be in its own section.

In this way then the void would be disproved also through the things that have already been said, and since those who suggested it thought that it was also the cause of motion, and a cause as being a space in which there is movement, he finally wants to show from that that the void in no way and nowhere is the cause of motion in any of the senses of cause.

In connexion with the question whether it exists we must understand what its name means. After he had set out the arguments on both sides, he turns finally to the search for the truth, and says that to learn whether it exists, that is, whether it is or is not, we must learn what the name of the void means, whether one thing or several, and what these are.

> **213b31** The reason for this is that they[40] think that what is, is body, [and every body is in a place, and void is a place in which there is nothing, and hold that while every body is in place, void is place in which there is no body, so that if anywhere there is no body, there is nothing there. Again, they think that all body is tangible and such is what would have weight or lightness. It follows therefore from a logical argument that void is this in which there is nothing heavy or light. This then, as we also said before, follows from a logical argument.]

For this reason, he says, they say that void is place in which there is nothing, because they say that all being is body. If then void is that in which there is no body, and all being is body, [void is] that in which there is nothing. It comes from a logical argument then that void is that in which there is nothing heavy or light. For if void is place in which there is nothing, and all being is body, and all body is tangible, and the tangible is all that which has weight or lightness, it follows that void is place in which there is nothing heavy or light. He did well in saying 'from a logical argument'. For they did not straightway say this, but they say void is that in which there is no body, and since every body is tangible, and they say tangible is what has weight or lightness, from the things said by them he concluded that it follows for them, from what they say, that void is what does not have anything heavy or light.

214a4 It is absurd if a point is void.

Saying that they say that void is place in which there is nothing heavy or light, he tests the definition by saying that on this account the point also would be void. For the definition fits it also: for there is nothing in it either heavy or light.

214a5 For it [i.e. void] must be place in which there is an extension between tangible body.

This is instead of 'it should have been defined like this, that place is an extension between tangible body'. Tangible is what has weight or lightness. So that place would be an extension receptive of heavy and light, or an extension of heavy and light.

214a6 But then void seems to be spoken of in one way as what is not full of body perceptible by touch. [Perceptible by touch is what has weight or lightness.]

The full has some things in it. And if it is full, how is it void? 'Full' is instead of 'It has been filled and is receptive of body perceptible by touch.'

214a9 Hence someone may ask, what would they say if the extension had colour or sound [would it be void or not?] 621,1

That even if they were to define it like this, even so they will not escape absurdity. For if, he says, this extension had colour or sound without weight and lightness, what will they say? That an extension like this is void, or not? The suggestion is not impossible. For the heaven is a body like this. For it is coloured and, according to the Pythagoreans[41] makes a sound as it moves, but it lacks weight and lightness.

214a10 Or it is clear that if it could receive a tangible body, it is void, and if not, not.

They will say, he says, in reply to this argument, that if you say that void is this, what is separated from tangible body, and to say the same [that it is separated from] heavy and light, even if it has some other body, it is clear that it is void (for we say that that is full that has weight or lightness), but if you say that void is that in which there is nothing at all, what is like this[42] is not void. We state the consequence for each: either the definition of void is discredited, or the same thing will be both full and void.

> **214a11** Another sense, [void is that] in which there is nothing,[43] and no bodily substance. [Hence some say that the void is the matter[44] of the body (who also [say that] place is this same thing), but what they say is not satisfactory; for matter is not separable from things, but they are looking for the void as something separable.]

That there is another sense in which they define the void, as an extension in which there is nothing, neither body nor anything bodily, like quality or potentiality or any other of the natural things. These people, he says, say that matter is both place and void, since it is actually none of the natural bodies, and clearly not first matter,[45] but that which has quantity. But it has often been shown that matter is not the same as place or the void.

622,1 > **214a16** Since a definition has been given about place, and it is necessary for the void to be place, if it exists,[46] lacking in body, [and it has been stated how place is and how it is not.]

From that he shows finally that the void 'does not exist neither separable nor inseparable'. First he says that with the arguments with which we have shown that place is not extension,[47] with these we show that the void also is not [extension]. For this place does not differ from the void, unless in thought alone and relations, since at least in their substrate they are the same. And we showed,[48] he says, that place, as limit, exists, but as some extension which in itself is different from the bodies that come to be in it, it does not exist. Hence also it would not be the void.

214a18 It is obvious that a void like this does not exist, neither separable nor inseparable.

On the one hand separable, as itself in itself entirely lacking body, as some used to say what is outside heaven,[49] and on the other inseparable, either as what is divided up in bodies, as those who supposed that atoms exist said, with bodies arising from its entanglement[50] with the atoms, or what is more, (he said) that there is not even a void like this, that on the one hand has its own nature, being an extension spread out in three directions as something other than bodies, and on the other is always full. But if we have shown that place is like this: with three extensions in itself, and the void is the same as place in its substrate,[51] it is clear that the void that is inseparable like this exists, which in itself is never separated from body, but is always full;[52] it has however a different existence from the bodies which come to be in it.[53]

214a19 For the void is not body, but they prefer it[54] to be an extension between body.

Since the void, he says, is an extension of body, that is, receptive of body, and place is also such a thing, according to those who think that place is extension, and it has been shown that the extension in place can be neither separable nor inseparable, it is clear that neither would the void exist as separable nor as inseparable.

214a21 Hence the void too seems to be something, because place is also [something], and for the same reasons. [For movement in place comes to the aid of both those who say that place is something besides the bodies that occupy it, and those who say that void exists.]

Because, he says, place and the void are the same, it is clear[55] from what, and through what reasons, they came to the thought that place was extension, through the same reasons they came to the thought about the void. For motion in place was the cause of the idea for both the one group and the other. For those who said that place was extension from seeing at different times different bodies coming to be in the same thing, as in the amphora,[56] came to this thought, as did those who said that the void existed because there would be no motion in place if void did not exist. For body does not pass through body.

> **214a24** They think that the void is a cause of motion in this way, that it is that in which motion occurs. [This would be like what some say place is.]

That those who believe that void exists think it to be the cause of motion in this way, that it is that in which motion occurs, that is on which.[57] And from this it is clear that the void is such as they also say place is.

> **214a26** But there is no necessity that if there is motion, there is void.

After refuting the void by the same arguments as [he used about] place as extension also, he hurries on from there to show that the void in itself does not exist anywhere in any way, but before that he wants to disprove the arguments by which they tried to show that the void exists, and wants to show that they have no necessary force. They were four, the first from motion, the next from growth, one from new wine, and one from ashes. And meanwhile he shows that the argument from motion in place has no necessary force. The argument was like this: if motion exists and a body does not pass through a
624,1 body, void exists: but the first, the second therefore too. For first, he says, even if some motion needs void, like that in place, surely not every motion needs void for it to come about. At any rate change[58] absolutely does not need place in addition. For what is changing, in its change, does not need place in addition. For it could be changed while being unmoved in place. But even if there were movement in place, it would not at once be necessary for there also to be void; for bodies could pass round one another by exchanging one another's places, and not needing in addition a void extension separate from them. And it is possible to see this in the case of movement in a

straight line, and is most clear in the revolutions of things that are continuous[59] as on the sphere of Aratus,[60] and with the water moving in the pot; for these things do not move through a void (for they occupy the same place in their own wholenesses[61]) but they move with their parts yielding their own places to one another. Why do I speak about things moving in a circle? For things moving in a straight line, as I have already said, move in this way not needing a void; for if I dip my finger in water and move it, all the things moving in the water simply move in place in a straight line, but they do not move through a void, nor are there separate voids in the water as big as the bodies of the fishes and those of the seals[62] (for that is obvious) but the water continually passes by the moving thing and each occupies the place of the other. For if the movement took place through solid and resistant bodies, it would reasonably have been supposed that moving things moved through a void, because the mutual exchange with them would not be easy, but if the movement is through fluid things, water I mean and air, it is not necessary to introduce the void because of movement. For with the medium through which the movement occurs being cut into and exchanging its own place with the moving thing, movement comes about in this way. From these arguments it is shown that it is not necessary, if there is motion, for there to be any separate void entirely devoid of body, but that it absolutely does not exist has not yet been shown from these. For what if, even if it were not through a void that motion occurred, there were otherwise some void separate by itself? On the one hand it has been shown sufficiently through these [arguments] that, as I said, it is possible, if there is motion, that there is no separate void, but on the other it has not been shown through these arguments that there is not a void inseparable in the way I said, which in its own definition is an extension void of all body, but is never apart from body, but is always filled with some body. But on the contrary it is absolutely necessary from what has been said for it to exist; for when we say that moving things do not require a void, but the parts pass round one another and yield their own places, what else are we to say they yield to one another than the extensions in which they are? Enough has been said about this in the discussions about place.

The second problem is from the condensation and rarefaction of bodies, of which an example was what happened with the new wine and the wineskins; for they said that when the body is condensed it goes into the voids existing in it.[63] He [Aristotle] therefore says that it is neither necessary nor true to say that condensation occurs in this way. For it is possible for bodies to be condensed not into the voids actually existing in them, but by pressing out and expelling the bodies with very fine parts in them, then to move into their place, which also happens with the wine;[64] for the stuff[65] with fine parts included in it is forced out by the pressure of the wineskins, whether it is like air or actually something else, such as for example smoke, or like vapour: new wine is like that. Therefore the wine pressed by this[66] and contracted into a smaller volume is given room by the same jar along with the skin, which, before the skin, it[67] actually filled. And, otherwise, there is much that is vapourlike in the wine, as shows the foam, which, when stirred by the breath, takes in a lot of air inside, as show its bubbles; with the motion therefore prevented by the compression from the skin, the air that has been taken in is carried out, as has been said, and in this way the parts of the wine move into its place. And, otherwise, there is always mixed into the stuff with coarse parts some finer body, so that water is compressed when air is squeezed out, and air when fire, and earth when both water and fire. 'Hence there is no compression of fire' as Themistius says.[68] For it has nothing with finer parts which it will contain.

The third of the problems was: 'if there is increase, there is void'; 626,1 for body does not pass through body. What is increased increases with the nourishment passing into it and being assimilated to what increases. It is necessary, then, if body does not pass through body, for the thing that produces the increase to move into the voids of what is increasing. Aristotle having earlier taken 'increase' wrongly as absolutely all change from smaller to larger, opposes the argument like this, but shows later that the problem is not compelling, not even with 'increase' in its genuine sense. For first, he says, it is not necessary for all progress from smaller to larger to come about with something coming into what is increasing; for with the change from water to air it has gone from smaller to larger, but it does not need anything coming in and producing an increase. So that the increase

has not come about through a void; for this reason the void seemed
to be implied by increase because the increase came about with some
body coming in. If someone were to say 'but if the volume becomes
larger it occupies some void spaces from outside' he will not be able
to show that this is necessary from the argument. For just as this
water in changing into air has changed from a smaller to a larger 15
volume, so also some other air in changing into water has
contracted into a smaller volume. And in this way equalising and
giving back come about. Aristotle not only refutes with reason the
argument in the case of the genuine increase, as he does next, but
also with the change from a smaller to a larger volume which
happens in coming to be and passing away. For those people did 20
not enquire into the kind of increase, and wonder about how this
comes about (for nowadays there is no argument about this), but
in wanting to show that void exists they made use of the question
about increase, that if a body becomes larger from smaller, and it
is necessary that the progress of the [smaller] bodies to the larger
comes about with some body entering and being assimilated, it is
necessary that void exists. So for demonstrating that void exists 25
the kind of increase makes no difference, but only that there is just
some progress towards the larger, and that it is necessary for this
to come about through a void. So with regard to this thought he
used this reply.

And, otherwise, if they were just to call all increase simply the
progress to a larger volume, they would be refuted like this, but if 30
they were to speak of increase in its proper sense, he says these
things to that. He includes with the problem about increase the one
about water being poured on to the ashes, and answers both together 627,1
and says that the argument about increase and the water being
poured on to the ashes entangles itself with itself, that is, tangles and
overthrows and refutes itself through itself. For, he says, there is a
common problem for the natural philosophers in the argument about 5
increase, and there seemed to be four particular absurdities, [of
which] certainly one could not be avoided: either there is no increase,
or it is not of a body, or the increase does not come about by body, or
body passes through body. In trying to solve these problems those
people supposed that the void existed, but no more in this way did

they solve the problems, if they did not actually extend them. So that neither did they show that the void existed nor did they solve the problems. I say again from the beginning: the natural philosophers enquire whether there is increase or there is not. It is necessary to agree that increase exists from the obvious facts. But again therefore is the increase of body or bodiless? But the obvious facts bear witness that the things being increased are bodies. Is therefore what as matter makes the increase body or bodiless? But this too is very clear. For the nourishment is body. If, therefore, both there is increase and it is of body and by body, and what produces the increase is assimilated in the thing being increased as a whole, since in fact what is increased is increased as a whole through a whole, it is necessary for body to pass through body. But that is impossible.

So the argument about increase is insoluble from every point of view. He says that those wanting to solve these problems about increase therefore put forward an hypothesis that was unprovable, namely, that the void exists (for, they say, there are voids in the body and the nourishment moves into them, and produces the increase in this way) and they appear by this both to preserve the things belonging to increase, I mean that increase is both of body and by body, and to escape the absurdity that body passes through body. But this argument, he says, overthrows itself and does away with itself; for they think that if they suppose that increase comes about through a void, they save the argument about increase, as increase would not come about other than through a void, but he himself says the opposite, that if void existed and the increase came about through it, there would not be an increase at all. For if what is being increased is increased as a whole through a whole,[69] as the obvious facts have it, the nourishment ought to be assimilated to the whole, and the nourishment is assimilated to the body not otherwise, as they say, than by passing into the voids in it. The body ought therefore to be void as a whole through a whole. For in so far as it is not void, clearly the nourishment is not assimilated to it, so that it will also not be increased. But indeed what is being increased is not increased as a whole through a whole; the whole then ought to be void. But this is absurd. If it were not void as a whole through a whole, the incoming nourishment would only be a filling up of the voids and not an increase; it is impossible then for increase to come about through a void. So they did not prove that the void existed, but assumed it without a proof, and did not resolve the problems about increase, but rather extended them. The same argument also applies to the ashes; for since both, the ashes are seen as a whole through a whole to be moistened, and the water as a whole through a whole to be darkened, there is every necessity that all the water be void as a whole and all the ashes be void as a whole, at least if they pass into the voids in one another, and this is laughable. And if the water and the ashes pass into the voids in one another, the ashes into those in the water and the water into those in the ashes, why had they not originally moved much earlier into their own voids? For it is more probable that the water, being liquid, should pass into its own voids, than the ashes into those of the water.

> **214a27** The void is in no way a general cause of all motion, [for a reason which escaped Melissus: for what is full can alter.[70] But not even [is it a cause] of motion in place; for it is possible for [bodies] to pass by one another without there being any separate extension alongside the moving bodies; and this is obvious even in the rotations of continuous bodies, as it is also in those of liquids.]

Till now he is refuting the general and unspecified part of the statement; for they just said this, that if there is motion, there is a void; to think, therefore, he says, or to say, that the void is a cause of all motion is laughable; for alteration,[71] which is motion,[72] certainly does not require place or void in addition. 'For a reason which escaped Melissus.' For he too, assuming that the all was infinite, declared that it was also motionless (for it did not have anywhere to move to); it escaped him too therefore that altering[73] could occur without needing place, and in this way it would not have been changeless. Not only, he says, is it possible to alter without needing another place beside the one in which it is, but [it is also possible] to move in place, since they[74] seemed to be speaking about motion in place. And this is clear with the things that move in a circle (for they turn round in the same place), but it is also possible to move in a straight line where there is no void, for the moving thing and that in which the movement occurs change places.

> **214a32** It is also possible to become denser not into the void but with the things inside being expelled [like water pressing out the air …]

That neither does the second problem have any necessity, the one from the wine and the jars; for it is possible for the wine to be compressed not into the pre-existing voids in it, but into the places of the bodies with finer parts expelled from it. The word *ekpurênizein*[75] has been used metaphorically from '*purêns*' or olive stones, which are forced out by the pressure of the fingers. Strictly the gigartons [stones] of olives are called *purêns*, and from these many other things also.

> **214b1** And to increase not only with something coming in, but also by alteration, like air coming from water. [In general both the argument about increase, and that about the water being poured on the ashes get themselves into knots.]

He refutes the third argument from increase, and for the time, as I said, uses 'increase' in a more common sense, calling just every change to a greater volume increase.

> **214b5** Either it is not increased at all, or not by body. [Or it is possible for two bodies to be in the same place (they claim therefore to solve a common problem, but they do not show that there is a void).]

The problem of the natural philosophers, which they tried to solve when they postulated the void. This problem, he says, those who postulate the void claim to solve; they do not however show that there is a void. For it is not necessary, because there is increase, for there to be void, but we must enquire into the cause of this kind of increase. For in postulating the existence of the void they do not solve the problem about increase.

214b8 Or it is necessary for all body to be void, if it increases everywhere and increases through a void. [There is the same argument about the ashes. It is clear then that it is easier to refute it from the arguments they use to demonstrate that the void exists.] 630,1

That the problem is not solved by postulating a void; for if the increase came about through some void, it is necessary for the whole to be void, if at least it is assimilated to all the body. We will say the same things also about the ashes.

Chapter 8

214b12 That there is no void separated in this way, as some say, let us state again.

Having shown that the arguments that imply the void involve nothing necessary, he wants finally from there to do away with the very nature of the void. Since those who spoke of it said that the void was either separate or inseparable, and inseparable [means] either what is potentially and by its own definition void, but always full of body, as those who said that extension was place would say that it was void, or what was scattered among bodies, as the followers of Democritus[76] used to say, first he shows that there is no void that is separate and having no body. And since those who said that the void existed used to say it was the cause of motion (for there would not be motion if void did not exist) he shows first that in no way whatsoever can the void be a cause of motion, neither as efficient[77] nor as final nor as that in which nor as that through which, I mean that just as we say now that motion occurs in size, it could not in this way also come about through void.[78] And earlier he shows that it is not possible for the void to be the efficient cause of motion. For since the void is in every direction itself like itself, and has no difference from itself, but we see that things moving naturally move in different movements, (for fire moves naturally upwards, but earth downwards) it is clear that the same thing would not be the efficient cause of different, much less of opposite [movements]; for the opposites are causes of the opposite [movements]. For we now, calling the natural forces the efficient causes of movements, assign opposite 631,1 causes to opposites (for the light is the cause of the movement upwards, and the heavy [the cause] of the downwards) but if the void were the efficient cause of movement, there would not have been different movement; further, there would not have been movement at all. For where will it move the body placed in it? Probably everywhere; the void is alike in every direction. So that fire will move upwards and downwards at the same time, and [so will] the rest, but that is impossible. So that the void would rather be the cause of rest, not of motion. It is therefore impossible for the void to be the efficient cause of motion. But neither would it be a cause of motion as final, I mean as the object of desire; for we will say the same things again. For the natural movements are different; their objects then are different; but the void is without difference in all directions; why then will fire move upwards more than downwards? For the goal and what is sought after is everywhere. Either therefore it will not move at all (for the void is like itself,[79] so that where anything is put in it, it has there its goal and object of desire), or if it does move, it will move from there in all directions; for why more up or down or in the other directions? For the object of desire, as I said, is everywhere. So that it will be torn apart. And there is the same story about the parts, and so on to infinity.

The same things, he says, that we state about the void, we have to say also about place against those who think that place is extension. For if the extension in place is the same as the void, what is said about the one [extension], can also be said about the other [void]. For if place is an extension without quality and bodiless, how will each thing be carried to its own place? For we, in saying that place is the limit of what surrounds it, since we say that it is the limit of a natural body, suppose that there are in it natural powers and peculiarities, and for this reason each thing moves to its own place as cognate, but if place were an extension without quality in itself, why will it move more like this than like this? For there is no difference by which the one thing will be moved in this way, the other in that way. And how among places the one will be up, the other down? For the extension has no difference. For as it is, since the bodies whose limits are places differ, and their limits differ with their cognates which they include[80] (for the limits are not in themselves places, like mathematical surfaces, but the surfaces on a natural body which are inseparable) but those who speak of the extension want it at once to be separate from the natural and simple bodies; for nothing will differentiate the extension that has received water from the one that has received fire.

But I would say in reply to these things on the one hand that it has been shown adequately in what came before that there is an extension other than the natural bodies, essentially separate from them, which will receive them, but in reply to the things now stated I say that above all it is not necessary for the place that will receive each of the bodies to have some power and quality. For neither is it the case that since it is good for each thing to be like this or like this, that one ought to put some power in that in which it is its nature to be. For it is not the case that since the body of the heaven is arranged in a circle, and that is good for it, Aristotle himself at least would say that the thing[81] in a circle had some power, desiring which the heavenly body is in a circle, at least because he does not even say that the thing in a circle is a place.[82] I say therefore that it is not because this part of the void, or at any rate extension, has some power that the heavenly body has occupied this part of it, like what is in a circle and outermost, but because it is the nature of the heaven to include everything in a circle and have everything within it; since then this is its nature, it is reasonable that it occupies the part of space that is like this. It follows then that it is not through the special power of the extension, but because the heaven is like this, I mean that it is inclusive of everything, that it occupies the outermost share of the extension. As therefore in the case of the creation of animals each different part takes hold of some different part of place – whether, according to Aristotle, of the limit of what surrounds it, or of the extension, as it seems to me and reason has demonstrated, like the head taking hold of its part of the surface of the air, even if it has not actually been divided off, and the hands of another, and the feet and so on; nobody says that it is because the limit of the air which touches the head has some power other than that which touches the feet or the hands, in such a way that it is with desire for it that the head has occupied this, and another part some other limit of the air, but that the head is naturally such as to overtop the rest of the body, and one of the parts is arranged in one way [and another in another] because this is good for the animal, because of this it came about that the head took hold of this, and other things of other parts, with the air lacking differences towards itself – I say that this also [applies] to the whole (for the universe is a big animal) since it was good for the union of the whole universe for the bodies that fill up the universe to have this kind of relationship to one another, each of the bodies with reason having as a natural urge the desire for such a relation to the rest, since this happens to it when it occupies this part of the extension, with reason it desires this, not because that has some power, but because both it is good for the all, and for each [individual] being and faring well are better when it is lower than this and is higher than that.[83] Therefore it desires its extension in place not for itself but through its relationship to the rest.

As therefore, as Aristotle himself says,[84] if the house were a natural thing, it would not have come about in any other way than it is now by craft, and it comes about with the roof being above and the foundations being underneath, and with the walls having the middle position, it is clear that even if it were a natural thing, naturally the roof would have moved to this part of the air, and the rest as it now is. Just therefore as now it is not because the air has some difference towards itself that one part of the house has occupied one part of the air [and another another], but because the house then has what is good for it when the parts have this kind of relation to one another, so also both with the parts of animals and with the greatest animal, the universe, the extension of the universe, which is both space and place of the all, has no difference itself towards itself. But by the fact that the bodies in the universe have taken up this relationship to one another, the differentiation of up and down has come about, and what takes up the most central relation we call down, and what the outermost and surrounding up. From these we have also given names to places, the part of the extension which has occupied the middle down, and what has occupied the outside up. Each thing therefore desires its cognate, as he himself indeed said,[85] when it has been torn away from it. It is therefore carried preferentially towards this, but in being carried towards this it is

also carried towards the extension which it has occupied, and its cognate wholeness has occupied this extension, as has been said, 634,1
through its relation to the all.

> **214b13** For if there is some natural motion [like upwards for fire, and downwards for earth, and towards the middle, it is clear that the void would not be a cause of motion. Of what then would the void be a cause? For it is of motion in place that it seems to be a cause, and it is not [a cause] of this.]

That the void is not the efficient cause of motion. If each thing, he says, has some natural motion of its own which differs from the rest, the void would not be the efficient cause of motion; for the causes of different movements must also be different, but the void lacks differences, so that the void would not be a cause of natural movements. Of what movement then would it be a cause? For if it has been shown not to be [a cause] of [the movement] of which it seems most to be, I mean of that in place, it is hardly a cause of another movement. So that the void is a cause of no movement.

> **214b17** Again, if it is something like a place devoid of body, when it is void, where will the body put in it go? For it is not into the void as a whole.

This argument could not be distinguished from the one before it in any other way than that the earlier one is directed against the suggestion that the void is the efficient cause, but this [is] against what supposes that it is the final cause. A distinction like this is reasonable; for he said above that since each thing has its own natural movement, the void, having no differences, would not be a cause of the different movements, but now he says, with the void having no qualities, where will what has been put in it go? What is carried towards something in a defined direction is carried as to a goal and what is reached out for. And it is reasonable for him, having set out to show that the void is not a cause of movement, to show that it is not a cause in any of the senses of cause.

635,1 **214b19** There is the same argument in reply to those who think that place is something separate, into which things are carried. [For how will what is put in move or stay still?]

Just as he related the arguments about place to the void, so also with good reason he relates those about the void to place. For he has already said[86] that place and the void are the same in their substrate; for the void is place devoid of body, according to those who think that place is extension. About place then he says, if place were extension, let us enquire, since we see that each thing desires its own place, light things up and heavy ones down, how, if the extension is without differences, are things carried to different places, one to one place and another to another. For why does fire desire upwards rather than downwards or right and left? For it should [desire that], if place is extension, and that is without differences.[87] And how in general will the extension have up and down if it is without differences? For up and down are different, or rather opposites.

214b22 [About up and down] and about the void the same argument will be appropriate, [with reason. For those who say that it exists make the void place …]

Having said that the same argument which has been made about the void will also be appropriate to place, if it were said to be extension, he turns the things said about place back about the void, in order to display their kinship. And he brings in the reason, that place and the void are the same. So that they are both introduced and done away with at the same time.

214b24 And in what sense will it be either in place or in the void? For it does not come about,[88] when a body is put, as a whole, in what is separate and permanent;[89] for the part, if it is not placed separately, will not be in a place, but in the whole. [Again, if not place, nor will it be void.]

He has already said the same thing also in his discussion about place, but there very confusedly and unclearly, but here more clearly. For from there that passage too has been considered, where he said as follows:[90] 'If there were some extension that was natural and remaining in the same position, places would have been infinite; for if the water and the air changed places all the parts in the whole will make the same thing, which is all the water in the vessel.' As I said then,[91] we have worked out the explanation of this passage from what was said there. What he says is this: it would be impossible, he says, for anything to have been put in the void or in place, if it were extension. For as it is the common conception is that the parts of continuous things are not in themselves in place, but the whole is in place; for the part does not come to be in place in itself, unless it has been divided from the whole, but it is in place as in a whole, but incidentally, because the whole and the parts are said to be in place. Since this is our conception of 'in place', that is not the case when some body has been put in a place separated and with extensions, I mean with the parts of it in itself not being in place; for with the extensions moving through themselves, just as the whole body has occupied the whole extension, so also each of its parts has occupied a part of the extension, and it is surrounded in itself by that, and, as he said earlier, if quantities are divided to infinity it will follow that both place and what is actually in place has been divided to infinity. This, he says, is impossible; for it is not possible for the part in itself to be in place, unless it has been cut off from the whole and in this way been made separate in itself. So, if this is impossible it is therefore impossible for a body to have been put in place or in the void; for both are the same, as has been said.

> **214b28** To those who say a void[92] is necessary if there is to be motion, rather the opposite is true, if anyone looks into it, [for it not to be possible for even one thing to move, if there is a void; for just as those saying that because of its all being alike the earth is motionless,[93] so also it is necessary for it to be motionless in the void; for there is not anywhere towards which it will be moved more, or less. For as it is void, it has no differentiation. Then that[94] all motion is either by force or natural.]

Having shown that the void is not a cause of motion either as efficient

or as final, he wants finally to show that neither can the void be a cause of motion as that through which, like the instrumental,[95] but before that he takes up the things that have been said and shows what follows from these for their hypotheses. For to those who say that the void is a cause of motion, he says, the opposite of what they say is the case; for they say 'if there is motion, necessarily there is void', but from what has been said the very opposite is shown, that if
637,1 there were void, there would not be motion. For, he says, just as some say that the earth is motionless because it is surrounded on all sides by what is alike (for it is not able to move anywhere because the whole of what is outside is alike; for why more this way than that way? For it is shut in by the equality and similarity on all sides alike), this is a consequence certainly even more reasonable for those who postulate the void; for if the void is alike on all sides there is nothing to which it would be moved more, or less. Having shown therefore that if there is a void it is impossible for there to be natural movement, he shows that neither is unnatural movement possible, and first he constructs his argument as following from what has been said, and then constructs an independent one that it is not possible for anything to be moved in a void unnaturally. For every unnatural movement, he says, is posterior to the natural, and the unnatural is a perversion of the natural, for there would not be an unnatural and violent movement that was not opposed to the natural. Since therefore it is impossible for natural movement to occur if there is a void, it is obvious that neither is unnatural possible; for in situations where the unnatural exists, the natural has much earlier preceded it. That it is impossible for natural motion to occur in the void he has already shown, and again takes up the argument. For if, he says, there is no differentiation in the infinite void, there would also not be different motions in it; for in that they say that the void is infinite, they do away with up and down and in general the centre and the outside, towards which is the natural motion for natural bodies, and in that it is void, they take away every differentiation in it. For, he says, just as there would be no differentiation in the nothing towards itself, so also in the void, and the void is not some thing, and is a privation. If therefore there is no differentiation in it, neither would there be different movements if there were void; but natural

movements are different: one of the two is necessary therefore, either for there not to be void, or if void exists, for there not to be natural movement. But if there is no natural movement if there is void, there would also be no unnatural [movement]. If therefore there cannot be either natural or unnatural movement in the void, it seems then that if a void exists there is no movement at all.

So therefore having shown that it follows from what has been said that it is not possible for unnatural motion to come about in the void, Aristotle shows independently that it is impossible for there to be motion in the void, like this: every unnatural movement comes about either with the source of movement present and[96] forced or with what gives the impetus being apart and not touching, as in the case of those who throw [stones] and shoot [arrows]. But the things being thrown are moved by force, or, he says, with the air changing places under the power of the thing being thrown (for the air is pushed by the thing being thrown, and, being dispersed into the sides, changes places with it backwards and so pushes it, and the change of places occurs so far until the power of the thing thrown slackens) – one kind of unnatural movement, therefore, seems to be like this, and another with the pushed air together with the thing thrown moving at a faster rate than is the natural speed of the pushed thing by which it is carried towards it own proper place, and in this way pushes it. For the air is easily moved, if it only gets a start, and it advances further preserving its given speed, and by its moving faster, as I said, than the natural speed of the object being carried along, pushing like this it moves unnaturally. Since therefore there are these two kinds of unnatural movement, in neither of these kinds can there be unnatural movement in the void. For the void cannot change places or be pushed; for the void is not a body. It is impossible therefore for there to be unnatural motion in the void, unless it is like things being carried, he says, and these are such as those which have the original mover accompanying them. He said this feigning ignorance: things are moved by being carried, like things on water, as when some chaff is carried on moving water; for this is moved unnaturally neither being pushed, nor by the interchange of places of the water, but by being carried on the water. This would not be called unnatural motion in the strict sense, but rather incidentally. It is not possible either for this movement to exist in the void; for the things being carried are carried on a body, not on a void, and the carrier must be difficult to separate from what is carried; for it is impossible to be carried on the void. And again, he says, at what point will a moving thing stop? For now we are able to say that the cause of its stopping is through the exhaustion of the tension of the pushed air, but where and by what will the thing moving unnaturally in the void be stopped? It is necessary then that either there is no motion at all, or 639,1 things are carried in all directions, unless something more powerful prevents it.

Aristotle shows[97] these things, that it was not possible either for forced and unnatural movement to occur, if there were void, but it seems to me that this argument has nothing necessary about it. For, first, in fact nothing has been proved sufficiently to satisfy our mind that one of the enumerated senses is a cause of unnatural and forced motion. And I have made a few remarks about this proposition in my notes on the eighth book of this work,[98] where[99] Aristotle primarily started an argument about these matters, how things moving unnaturally move, but also now, none the worse, he records shortly the unpersuasive things connected with this argument. For in the exchange of places either the air pushed from the front by the arrow or stone that has been thrown runs backwards and changes places with the arrow or stone, and in this way getting behind pushes it, and so on in turn, until the force of the thing thrown is exhausted, or what exchanges places is not what is pushed from the front, but that from the sides; for with the arrow pushed by the air originally pushed with it by the bowstring, the air from the sides changes place into the place of the arrow, and the air, also pushed by the air originally pushed, moves the arrow, and in this way will do the same thing again on the air changing places with the arrow and so on until the force of the movement given originally is exhausted. If therefore we say that the change of places occurs in the former way, I mean the air pushed from the front by the arrow changes places with it backwards and pushes it, someone would wonder why what will force the air which has been pushed forwards once, with nothing opposing it, to run backwards, clearly on the sides of the arrow and, coming backwards, to turn back and push the arrow. For it is necessary for there to be three movements; it must be pushed by the arrow forwards, then run backwards, then turn back in the same movement. However, the air moves easily, and when it gets an impetus to movement it goes furthest. How then is it that what is pushed by the arrow does not move according to the impetus it has been given, but running backwards as on an order bends back in the same way, and running backwards is not scattered into the yawning gulf, but aiming accurately at the notch of the arrow runs back to it and grasps it? These things are entirely unbelievable and are more like fictions. Then, the air in front which has been pushed by the arrow clearly makes some kind of movement, and the arrow also moves continuously. How then is it possible for the air pushed by the arrow to change places with the arrow and be in the place which the arrow had left? For before this can run backwards, both the air from the side of the arrow and that which falls behind coincide and instantaneously fill up the place left by the arrow through the force of the void, and above all that which moves behind along with the arrow. If anyone were to say that the air pushed by the arrow runs backwards and pushes the air that is changing places with the arrow, and in this way gets behind the arrow and pushes the arrow into the place of what has been pushed, it was above all therefore necessary that the arrow should not move continuously; for before the air changing places from the side was pushed, it is clear that the arrow did not move. For this air did not move it. And if this did in fact move it, what was the use of the one before running backwards? And how in general or by what did the air pushed forwards get the impetus for its movement backwards? For if it can push the air falling on it at all, as it has now pushed the air from the sides, why is it not at the beginning of the motion which it received from the arrow that it does this most, and pushes the air falling continuously on it in front, but makes double and triple pipes[100] outside the cause of movement? So there were many things to say and to refute this inventive hypothesis. But these things are enough for our project.

In reply to the second hypothesis, which said that the air from the sides changes places, and this is pushed by the air which was originally pushed, I say that if the air from the sides is actually unmoved and changes place with the arrow, and gets into its place, it is much more probable that the air pushing the arrow from behind and moving briskly in such a way that it stirs up another exchange of places from the sides, should also continuously change places with the arrow, and push it, until the start of the movement originally given to it expires.

These remarks on the one hand are directed at the argument which says that things moved by force are moved by the changing of places by the air, but on the other hand, at the next argument, which says that the air is pushed and takes an impetus of movement, and moves more briskly than the natural speed of the thrower, pushes it and so pushes it and accompanies it until the kinetic power given to it expires, which has more plausibility, there are the following things to be said, but the argument that will have been said will also be appropriate to the one about the changing of places. One must ask those who say these things: when someone throws a stone by force, is it by pushing the air behind the stone like this that he forces the stone into an unnatural movement, or does the one who pushes give some kinetic power to the stone?[101] If then he gives no power to the stone, but only by pushing the air he thus moves the stone, or the bowstring the arrow, what was the point of the hand touching the stone or the notch of the arrow the string? For it was possible by not actually touching these, but as it were standing the arrow on a piece of wood or on some fine line, and likewise the stone, with ten thousand machines to move a great amount of air, and it is clear that by how much the greater amount, and with the greater force, the air is moved, by so much the more it ought to push and force a thing out. But as it is neither would you stand the arrow or the stone on a line or a mark actually without width, and move all the air behind with all power, nor would the arrow be moved the distance of a cubit. If therefore the air moved with greater power were not moved, it is clear that neither with those who throw stones nor with those who shoot arrows is this air pushed by the hand or the string the mover. For why should this happen more with the thrower touching the thing thrown or not? And again, if the arrow in touching is in contact with the string, or the hand with the
642,1 stone, and there is nothing in between, what would the air moved from behind be? And if the air from the sides is moved, why towards what is being thrown? For it falls out of that.[102]

From these and many other arguments it is possible to understand that it is impossible for things moved by force to move in this way, but it is necessary for some incorporeal power to be given by the thrower to the thrown thing, and that either the air pushed does not contribute at all to the movement or very little. If things thrown by force are moved in this way, it is surely clear that even if someone were to throw an arrow or a stone by force and unnaturally in a void, the same thing will happen much more, and it will not need something pushing from outside. Surely this argument will not be more difficult, being supported by what is obvious (I mean that some incorporeal kinetic force is given by the thrower to what is thrown, from which the thrower ought to touch the object being thrown) than that some energies arrive on the eyes from the things seen, as it seems to Aristotle.[103] For we see from colours some energies arriving incorporeally and colouring the solid bodies existing in front of them, when a ray of the sun strikes through the colours, as is clear to see when a sunray strikes through the coloured glasses; for on whatever solid body the ray falling through the glass falls, that it colours in a way similar to the colour through which it fell.[104] It is clear therefore that some energies come bodilessly from some things to others. Next, if what is thrown is the stone or the arrow, and what puts it in motion or gives it power is the man, what prevents [our] throwing something with what is between being void? For if, as it is, what is in between is body and that hinders the movements of bodies, and although things that move need to divide what is in between they yet move, what prevents [our] throwing an arrow or a stone or something like that if what is in between and the thrower and the thing thrown and the space are void?

215a2 It is necessary that if there is forced [movement], there is also natural; for the forced is unnatural. [But the unnatural is posterior to the natural, so that if there were no natural motion for each of the natural bodies, there will also be none of the other motions. But how will there be natural motion if there is no difference in the void or the infinite? For *qua* infinite there will be nothing up or down or central, and *qua* void the up and the down do not differ (for just as there is no difference in the nothing, so also in the void) ...]

Since he divided movement into the forced and the natural, and the forced helps him in his proof, for this reason he says that if there is forced movement, there is also natural. For what is by force is
643,1 nothing other than unnatural. If therefore there is unnatural movement when there is a void, natural movement must take precedence. But natural movement does not exist [in a void], and so therefore neither does unnatural. Next he shows how natural movement does not exist [in a void], but since no necessary argument has demonstrated that there is no natural movement in the void, the unnatural would not in this way be done away with. Certainly I do not say that what moving things move through is actually void, but that even if there were void nothing would have prevented movement.

> **215a11** Since the void does not exist as something, it would seem to be privation.[105]

Not that it does not exist in any way at all, but like matter and privation it is said to be what is not because it is not actually any of the natural bodies, and thus the void too would be spoken of as privation and what is not by the fact that none of the natural bodies exist in it.

> **215a11** Natural motion has differences …

For some things move upwards naturally, others downwards; so that natural movements differ.

> **215a12** … so that natural things[106] will be different. [So either there is no natural motion of anything anywhere or if there is there will be no void.]

That is, things moving naturally. And if things moving naturally differ, but in the void there is nothing different towards which they would move, one of two things is necessary, either motion does not happen in a void, or nothing moves naturally.

> **215a14** Again as it is [things thrown] move without the thing that pushes them touching them[107] [either through a change of places, as some say,[108] or through the pushed air pushing in a faster motion than the speed of the pushed thing at which it is carried to its own place. But in the void none of these things apply, and there will not be being carried except as what floats.]

From here he produces an independent argument that unnatural motion would not occur in the void. The words 'with what pushes not touching' are instead of 'not touching until it moves' since in pushing it clearly touches the arrow or the stone. Since the hand does not include any air between itself and the stone, in this way it pushes the
stone by pushing the air, and similarly nor does the string have any 644,1
air between itself and the arrow, but the surfaces fit one another, that of what throws and that of what is thrown, is it that before touching the air behind the thing being thrown, and so pushing it, there comes about some movement in the stone immediately from the hand, or in the arrow from the string, or not? If then there is no thrust from the hand or the string in the stone's being thrust to the place not natural to it, it ought not to be at some time at a distance from the hand;[109] for it is neither carried by its natural rush to what is not natural, nor can the hand immediately thrust it. But as it is, it is thrust with no air left between; so that if there is no air in between, the primary thruster is the hand or the string, if finally, as they say, the air also receives the thrust. And if it is possible for the thrust to be produced wholly and immediately by the hand, what stops this same thing coming about also in a void? For to me there is no difference whether it comes about more or less through a void or through a body.

> **215a19** Again no one would be able to say why, when it has been started, it stops [anywhere. For why more here than there? So that either it will stay still, or it will be necessary for it to be carried to infinity, unless something stronger prevents it.]

This is another argument, that nothing moves unnaturally in the void. I say that just as you who think that the cause of unnatural movement is the thrust of the air, say that it moves so far until the kinetic power given to the air from what originally pushed it expires, in this way clearly even if something were to move unnaturally, in the void it would move so far until the kinetic power given to it by the original thruster was exhausted.

> **215a22** Again, as it is, it seems that things are carried into the void because it yields,[110] but in the void this sort of thing is entirely uniform, [so that they will be carried in all directions.]

This is another argument from the things said by those who suppose that the void is dispersed in bodies; these are the followers of Democritus. For now, he says, you give the cause of things moving
645,1 more in air, less in water, and not at all in earth, that all things arise from the combination of the void and the atoms, but the combination is not uniform in all things, but the voids in the bodies[111] are greater or smaller. For this reason movement through air is faster, because the voids in air have large parts; for the atoms of the air pushed by the moving thing pass easily into the voids and give a passage to the moving thing. In water the movement is slower. For the voids are smaller, and the compression into them takes more time. In earth, because the voids have very small parts, movement does not occur, for the atoms cannot be compressed into them. If therefore, he says, with these things motion comes about like this, if the whole were void, where, he says, will they go? For it is not that one part of the void is more yielding than another. But everywhere nothing is a hindrance. So that things will be carried in all directions alike. For as it is, in the direction in which the atoms yield to what is pushing, the things moving move up or down or wherever, but in the void which is yielding in every direction alike, if one ought to call the void yielding, but rather not hindering anything in any way, if is reasonable that the moving thing be carried in every direction and no more in this way than in that way. This is irrational and impossible. And if the larger the parts in the void are, by so much the faster is the movement, if then the whole were void, movement in it will be instantaneous.

It is possible to take this argument together with those above, because now they would equally give this explanation of why the things moving unnaturally are not carried through everything in the same way, but if the whole were void, what would they say? For things would be carried either instantaneously or everywhere. Rather, through this argument it has been said that neither can movement occur through a void as a medium. For neither will the faster and slower be in it, nor the more in this way than in that way. For everywhere is similar. So that it should be ranked more with those [arguments] that come next, not with those before.

> **215a24** Again, what is said is obvious from the following: for we see a body with a fixed weight being carried faster for two reasons [either by a difference in the medium …]

Having shown that the void is a cause of movement neither as efficient nor as final, he wants to show finally that it is not either a cause of movement as its medium, such as that the void is a cause 646,1
because bodies are carried through it, as now the movement of bodies comes about through air or through water. For if it had been shown that not even like this is it a cause of movement, nor is it possible, if void exists, for natural movement to occur through it; and it has been shown that nor is unnatural movement possible; it appears therefore that if void exists it is not possible for movement to occur. But indeed movement does occur: there is therefore no void, which was the original proposition to be proven. He shows that it is impossible for there to be movement in a void because of unequal[112] movement. For there exists some movement that is faster and some that is slower. We see then, he says, from the facts themselves, that unequal movements come from two causes, either with the medium differing, (for the same weight moving through water will be moved more slowly, but through air more quickly), and if again the medium through which things move were the same, but were not in the same state, the movement becomes unequal. For just as a thing moved more slowly through water because that is more difficult to divide than air because of its density and its spending time in the act of dividing, so if the medium through which it moves is the same, like the air, but it is carried now through what is still, but at another time

through what is moving in the opposite direction, it moves more quickly through what is still. For with its being thrust back by it, it is hindered from going forwards; for the division becomes more difficult because the moving air glides off and because of the opposing thrust. In this way the motion becomes unequal because, while the moving thing is the same, that through which it moves is not the same, and if on the other hand the medium is the same but the moving thing is not, in this way too the movement becomes unequal. For the pound weight[113] in the same air is carried down faster than the half-pound, for the greater force divides it more quickly. But if there is the same force, like that of two pounds, but they differ in their shapes, and the one is spherical and the other flat, the spherical will be carried down through the same air more quickly than the flat; for the more angles it has, the faster it makes the division, and what has many angles, as it gets more, will approach the angleless, I mean the spherical. Fastest then of all the spherical shape is carried down (for it makes the cut by a track[114]) but the flat, being held up by more air, makes its cut more slowly, like the blunt edges of knives. Hence the more like a sphere something is, the easier it is to roll, like the cylinder and the shapes with many angles. Hence the sphere is the easiest to roll of all (for it touches the surface at a point), and the cube is the most difficult of all to roll; for it occupies a large share of the surface. So also with things being carried downwards, the spherical make their movement faster because they are held up by a small amount of air, and the small, if it is of the same kind, is more easily divided than the larger.

The causes of the unequal motion are these, but it is impossible, he says, if the motion happened in a void, to give the reason why unequal motions occur; for it is necessary for everything to move at the same rate in a void, but rather it is not possible at all for motion to occur in a void. He shows this in this way: he takes two distances[115] equal in length like that of a stade,[116] air and water, and some one thing moving through both, or two with equal force and equal shapes (for there is no difference) like two spheres of gold of a pound in weight; if then one were moved through the air and the other through water, if the air has finer parts than the water, clearly the one will get through the stade-long air more quickly than the other the water.

The times of the movements will have the same ratio[117] to one another as the ratio the air has to the water in density and fineness. If the air had parts finer than the water by a twofold ratio, necessarily the times of the movements would have the same ratio vis-à-vis one another; for the time in which a thing moves through water will be double the time in which it moves through air, and the finer the parts of the medium through which the movement takes place, by so much is the time of the movement reduced. Since these things are so, he therefore shows that it is impossible for movement to occur through a void in this way: he shows it with letters,[118] and he calls the moving thing A, the air D, the water B, the time of the movement through the water C, that of it through the air E, and he takes a void equal to the water and the air, which he calls F. If then it is possible for movement to occur through a void, since all motion is in time, let the A have moved in a certain time through the void F, 648,1
and let, he says, the time in which it moves through the void be G.

Since then there is some ratio between every definite length of time and every [other] definite length, it is necessary that the time in which it moved through the void have some ratio to the time in which it moved through a substance.[119] If the time has a ratio to the time, and just as the times have it to one another, similarly the things through which the movements occur must have the same, and contrariwise the ratio the things through which the movement occurs have, this the times must necessarily have; clearly the ratio which the time of the movement through the void has to the time of the movement through what is full, like water, this ratio the void has to what is full, for instance if the time in which there was movement through the water was tenfold that through the void, the water ought to be ten times denser than the void, and the void obviously with parts ten times finer than water. But it is impossible for the void to have any ratio to body. And he constructs this argument clearly. For what exceeds something, he says, includes the whole of what it exceeds and the excess by which it exceeds it. Thus four exceed three by a unit; four then have both the three which they exceed, and the excess by which they exceed them, I mean a unit. Just as a number does not exceed nothing or have any ratio to nothing (for if a number exceeds nothing, it is clearly composed of the excess and that which

is exceeded, so that the tetrad would be composed of the four; for it is by these that it exceeds nothing; and of the nothing; for it exceeds that. What then is the difference between the tetrad composed like this and what composes it?) – so just as it is stupid to say that a number has some ratio to nothing and exceeds it, thus clearly and like this it would be ridiculous to say that some body has a ratio to the void and exceeds it in density; for it is composed of the void and the excess, so that water is composed of the void and water. But that is laughable; it is therefore impossible for the void to have a ratio to body. If the things through which movement occurs have no ratio to one another, nor clearly do the movements, nor will the times of the
649,1 movements have a ratio to one another; for it has been shown that as the things through which the movements occur are to one another, so are the times, and contrariwise as are the times, so are the things through which. If therefore the void has no ratio to body, nor does the time in which movement occurs through the void [have a ratio] to the time in which it occurs through body; in no time then will there be movement through the void. For there is some ratio between every definite time and every definite time; therefore there is no movement in time through the void. So neither will anything be moved through the void; for every movement is in time; for there is no timeless movement.

This then is his first argument, and the second is like this: if something moves in time through the void it will clearly move in less time than it moved in in what is full, like the air, with the distances[120] being the same, as has been said. For see, if it moved in two hours in the air, it would clearly move in less time through the void, say in half an hour. But in this half [an hour] in which it moved through all the void, it will clearly also move proportionately through some part of the air, and he calls that part of the air H. It will follow therefore that the same thing in the same time will move through a void and what is full,[121] which is absurd, with the earlier [argument] also remaining absurd, I mean that the void has some ratio to the full, for the ratio that the part of the air has to the whole of the air, this the whole void has to the whole air (since the time taken to pass through the part of the air and that to pass through the whole of the air have the same ratio [as has the part of the air][122] to the whole of the air) and again the ratio that have as full to full the part of the air to the whole, because the times for passing through the void and for passing through the part of the air also have the same ratio to passing through the whole air. But he himself leaves out this absurdity, as being obvious, I mean the earlier one that the same quantity moves through the void and the full in the same time; and lest anyone should say: 'But if in the same time it moved both through the full and through the void, but not the same distances, for we assumed that a fourth part of the air was moved through in the same as was the whole of the void, but the void is equivalent to the whole air: therefore four times the distance was moved through in the void in the same time, and this is nothing absurd, that the greater length, but with finer parts, moves over the same amount in an equal time as the smaller and denser', for this reason he next shows that of necessity if there is movement through a void, the same distance will have been covered in an equal time through the full and the void, which is impossible. For if, he says,[123] we think about some substance finer than the air with this ratio to it which had the time in the air, which was the E, to the time in the void, which was the G (let's say, if the E time were four times that of the G, it would be a substance with parts four times finer than air), let us place this substance in the void, with the distances obviously remaining the same. The A will move through the finest substance, which we placed in the void, in the same amount of time as that in which it moved through the part of the distance in the air, which was the H, and that was the fourth part of it. But earlier it moved in the same time in the void as now in the finest substance over the same distance as in the void, therefore in the same time the A will now cover the same length through a void as through something full, with them being equal in their distances, which is absurd. So, he says, we must do away with the hypothesis from which the absurdity followed, and this was the suggestion that motion occurred through a void. The reason, he says,[124] for the diversion[125] into absurdity is clear. For since every movement is in time and there is a ratio of every time to every time, if they are definite, it is clear that even if movement occurred through a void, it will both be in time and will have some ratio to the time of some movement: and as the times of the movements are to one another, so naturally will be the things through which the movements occur; therefore the void will have some ratio to the full. But it is impossible for the void to have a ratio to the full (for it exceeds all ratio) and from that follows the absurdity: for if it is taken that movement occurs in a void it follows that the void also has some ratio to something full, and that is impossible.

Up to this point Aristotle's argument proceeds from the inequality of the speed of moving things compared with that through which their movement occurs, arguing that it would not be possible, if there were a void, for motion to occur through it, but anyone who has the goal of arriving at the truth from all sides, let him look round at how great the power [of these arguments] is, lest through the harshness and obscurity of the arguments he misses his goal. It is better
651,1 perhaps first to go through the whole argument about the void, and then to take up each of the arguments from the beginning and enquire what truth or falsity is in it, respecting nothing, and not putting the reputation of the man before the truth.

215a26 ... as through water or earth ...

He took the earth as an example, not as if movement occurred through it. He says that earth is a muddy[126] space, and then he goes over to clearer examples and takes air and water.

215a27 ... or because the thing being carried differs, if other aspects are the same, in the excess of its weight or its lightness.

After speaking of the one cause of the unequal movement, I mean differences in the medium, he talks now about the rest, I mean the differences in the things being carried; for if the medium is the same, but the things being carried differ in weight and lightness, (I mean by lightness the lower weight, as if the one were a pound weight, the other half a pound), and if they were carried through the same air, they would not be carried in equal times. The words 'if other aspects are the same': for example, if they do not differ in their shapes; for it is possible for the heavier to move in an equal time as the lighter over the same distance, if the heavier were flat but the lighter spherical.

It is possible to understand his words like this. In 'the thing being carried differs'; we understand the 'differs' either in size or in shape, and the 'if the other aspects are the same' as, if they differ in size, both the shape and the medium are the same, and if in shape, both the size and the medium are the same; for the movements would be unequal like this.

> **215a29** The medium through which it is carried is a cause, 652,1
> because it resists most what is carried against it, but then what is staying still [and more what is not easily divided; and such is what is denser.]

It resists the movement either by advancing in a direction the opposite to the things moving (as when either something is carried through air which is being moved in the opposite direction, or through water with the flow going in the opposite way to what is moving, as if someone were swimming in a river, or a boat were to be carried against the flow), and[127] the medium not moving, he says, but staying still, it resists being divided. And if the dividing also resists, what is most difficult to divide will resist more.

> **215a31** That on which A will be carried through the B in the time in which C.

After saying how many and what are the causes of the unequal movement, he wants to show how the medium is the cause of unequal movement through letters (for he will speak after this about the thing that is moved), and he calls the thing moved A. Let this therefore be moved through the distance B, and let this be water, in a certain time, like the C.

> **215b1** Through the D having fine parts, if the length of the B is equal to that of the D, with the substance resisting proportionately. [For let the B be water, and the D air; and by how much the finer air is than water, and less corporeal, so much the faster the A will move through the D than through the B. Let

there be the same ratio as air has to water, the speed to the speed, so that if it is twice as fine …]

The thought of the text and its arrangement are this: let, he says, that on which is the A, which is the moving thing, move through the D, a substance with fine parts, and move the same length as is that of the B, in the D, and move 'proportionately to the resistant [substance]', in a denser thing over a longer time, in a finer one over
653,1 a lesser. He calls the B water, and the D air. 'By how much the finer air is than water, and the less corporeal, so much the faster the A will move through the D than through the B.' For if the medium is resistant to being divided, by how much a thing is more easily divided, by so much the faster what is carried will be carried through it; what is finer is easily divided, and air is finer than water and, as among substances,[128] less corporeal. So that by how much the air is finer or rarer than water, by so much is it also more easily divided. So that by so much faster will the A move through it than through water. What ratio therefore the magnitudes have to one another, this ratio the movements through them also have, and the times of the movements, for instance as if the air were twice as fine as the water, by a twofold ratio the movement through the air would be faster than that through the water. And the time of the movements will therefore inversely have the same ratio. For the time of the movement through the water will be double that of the movement through the air.

215b8 In twice the time that it traverses the B than it does the C[129] [and the time in which the D will be double that in which E.]

The B than it does the C, that is the distance which has the B than the C.

215b10 And also always by how much that through which it is carried is less corporeal and less resistant, but easily divisible, it will be carried the more quickly.

He has used the conjunction 'but'[130] instead of 'and', because they have the same meaning; if a thing is carried more quickly through what is finer, by how much something is less corporeal and more easily divided, and for that reason less resistant, it will be carried more quickly through that.[131]

> **215b12** But the void has no ratio by which it is surpassed by the body, just as the nothing has to number.

Having shown the relations of movements with regard to substances, he comes finally to our proposition that it is impossible for movement 654,1
to occur in a void. For if every movement is in time, the movement through a void would also be in time. There is some ratio between every time and every defined time (for a time is equal to a time, or half or double or one and a third or by some other ratio), and as are the times of the movements, so necessarily are the things through which the movement occurs; so the void too will have a ratio to something full, and that is impossible. If therefore it is impossible for the void to have any ratio to any substance, the time of the movement through it will also not have a ratio to the time of the movement through any substance. But that is impossible; for there is some ratio of every time to every defined time. So there is movement through the void in no time. But this again is impossible: for all movement is in time. So it is not possible for movement to occur through a void.

> **215b13** For if four surpasses three by one, but two by more [and by still more one than two, yet it will not have an amount[132] by which it surpasses the nothing; for it is necessary for what surpasses to be divided into the excess and what is exceeded, so that four [will be divided into] the amount of excess and the nothing.]

After saying that just as there is no ratio between the nothing and number, so there is none between the void and substance, he argues this itself through what follows, how the nothing has no ratio to substance, and he gives an example of things that have a ratio to one another. For, he says, four exceeds three by a unit, and two more than by a unit (for it exceeds two by a dyad) and it exceeds the unit again by a greater excess than the dyad (for it exceeds the unit by a triad), but it does not exceed the nothing by any excess. Why it does not exceed the nothing he adds in what follows, and we have already said fully.[133]

215b18 Hence neither does a line exceed a point, unless it is composed of points. [Likewise too the void cannot have any ratio to the full, so that neither can movement.]

Lest someone should say 'But you took as a poor example of the void what is not; for the void is not a thing that does not exist', for this reason he adds this, that not only does what is not have no ratio to what is, but also things that exist but are *sui generis*[134] do not have any ratio to themselves. For neither does a line have a ratio to a point, 655,1 since the line is not composed of points, nor has a line to a surface, nor the latter to a solid;[135] for a thousand points added on do not make the line bigger, nor the lines the surface, nor the surfaces the solid; for they are not composed of these. But cognate things are compared to one another and have a ratio; for a line has a ratio to a line, and a surface to a surface, and a solid to a solid. So that even if the void is one of the things that exist, since it is entirely different from bodies, or rather is opposite like a habit[136] to privation, (for the one is void, the other full), it would not have any ratio to a body in that it is void. For we are now looking for the ratios of composites, that air has a composition of this size, and water of twice that; but the void therefore has no such composition, so how is it possible for it to have some ratio and comparison to bodies? So that neither would the time of the movement through it have any ratio to any time of the movement through a body, and in this way neither would there be motion through a void, as has often been said.

215b21 But if it is carried through the very finest in so long a time over such a distance,[137] through the void it exceeds every ratio.

If, he says, it moved through the finest substance in such a time over such a distance, and continually with a finer substance the time of the movement gets smaller, obviously with all the other things remaining the same, if the movement occurred in no substance, but

in a void, the time through the void will have no ratio to that through the full, and it is the same to say that there will be timeless movement through the void, and it is the same to say that there will not be movement at all in a void. For as with the sphere of Aratus,[138] and even more that of the all, the signs of the zodiac in the largest circle move faster than the rest, and those in the smaller circles more slowly, and the proportion of the slowing down is that of the decrease in size of the circles in which they move – for whatever is the ratio of the smaller circles to the greater, to that extent do those in the smaller circles move more slowly than those in the larger, and to this extent the decrease in size of the circles and of the movement on them 656,1
progresses until the circles taken on the sphere are expended and we arrive at the poles, which are completely motionless; hence Aratus says these things about the Ram, which is in the largest circle, I mean that where day and night are equal[139]

> The paths of the Ram are the fastest.
> Which is pursued round the longest circles
> And runs no more slowly than the Bear Cynosura,[140]

who goes round in a narrow circle, but, however, the one moving in the greatest circle is brought back from the same [point] to the same at the same time as that moving in the smallest, by the smallness of its circle, with the movement also being diminished – so therefore also with our proposition, if, to the extent that the fineness of the substance increases, the time of the movement is lessened, with it progressing to the faster, if then there is no substance, but there is void, the whole of [its] time will have been reduced to nothing, and there will not be any movement.

215b22 Let the F[141] be a void, equal in size[142] to the B and D.

Setting out the ratio that arises in the case of substances when something moves through them, and saying, or rather showing, that the void will have no ratio to the full, on which it followed that also no movement could occur through it, and wanting to make it clearer he sets out again with letters the relationship between the movements and the things through which they occur, and he inquires if it is possible even in the void to preserve this relationship. Having earlier set up water and air he asks what happens when movement occurs in these, and now, taking these things as agreed, [he asks] what kind of relationship, if possible, they would have to the void, if something were to move in it. And having made it an assumption that movement occurs in it, he leads[143] the argument to an absurd conclusion. He assumes therefore the void being equal in size to the B and the D, that is, to air and water.

657,1 **215b23** If the A passes through [the void], and will move [in a certain time, let it be G, but in a lesser time than that of the E, this is the ratio that the void will have to the full. [But in the amount of time on which is the F, the A covers that part of the D on which is the G.]

If, he says, the same A that moved both through water and through air were to move through the void in a certain time, say in the G (for he calls the time in which it moved through the void G), it is necessary that this time should be shorter than the time in which it moved through the air, and this is the E. If, therefore, he says, it moves through the void in a shorter time than through the air, clearly the time in which it moves through the void, say the F, will have a certain ratio to the time in which it moves through the air, and this is the E; the ratio therefore which these have to one another, the full and the void will have this ratio.

215b26 But in the amount of time on which the F, the A traverses the H of the D.

The second argument. In the amount of time, he says, in which it traversed the void, and this is that on which the G, the A will move a certain part of the distance D, I mean that of the air, and let this be the distance on which is the H. Leaving out the absurdity that followed this one[144] he goes on to the last argument (this is that in the same time G the distance H was moved over both as full, i.e. as a part of the air, and as void, which was F),[145] that full and void will have the same ratio to one another, I mean the F and the H, which is absurd; for the ratio that the H has to the D, I mean the part of the air to the whole of the air, this too the F will have to the D, I mean the void to the whole of the air, since also the time of [traversing] the F and the H, which is the G, has the same ratio to the time of [traversing] the D, which is the E, as the ratio the times have to one another, this the mediums have too; the F and the H therefore have this to the D. The things therefore that have the same ratio to the same thing, also have one and the same ratio to one another; the F, however, which is the void, and the H which is full, have the same ratio to one another; but this is absurd, for the full to have the same ratio to the void as the void also has to the full; for they will be the same as one another by their association (for it is by this that the proportion now is), and as the H is becoming thinner, I mean the part of the air, it would have become what the void is.

> **215b27** It covers, if there is something different from air in fineness which is the F, in this proportion which the time which is E has [to that which is G.]

The third argument, in which it is shown that if something moves through a void, it will follow that it moves in an equal time over the same distance both full and void. The same F which before this was void, he now makes full of the finest substance, which has this ratio to the air that the G time, that of the void, has to the E time, that of the air.

> **215b30** For if the substance called F is so far finer than the D as the E is longer than the G, [inversely the thing A will cover with speed in the time as great as the G the F, if it is carried.]

If, he says, the substance placed in the void were by so much finer than the air as the time of the movement through the air exceeds the time of the movement through the void, inversely, he says, 'the thing A covers with a speed as great as the G, the F, if it is carried', that is, by the extent by which the time of the movement through the air exceeds that in the void, by so much contrariwise the A will cover more quickly in the F in the time G than through the D. 'Inversely', he said, because with regard to the times the time of [traversing] the air exceeded the time of [traversing] the void, I mean that the E exceeded the G (for the E was greater), and in the case of movement the movement in the time G is greater in speed by the same proportion as that in the E. He did well in adding 'if it is carried', that is, if it is at all suited by nature to move through that substance that has been put in the void and is of that kind as we suggested, the A will move in it by the same proportion.

659,1 **216a2** If therefore there is no substance in the F, still faster. But it was [assumed to do it] in the G; so that in an equal time it covers what is full and [what is] void; but that is impossible.

In his conclusion he has brought together two absurdities which follow from his hypothesis. For it had been shown that the A will have moved through the finest substance placed in the void in the time G, but it had been supposed from the very beginning that it had moved in the void over the same distance in the time G, so that in the same time this same A will have moved over the same distance F both full and void, which is impossible. This indeed he himself brought in later, but before this drew another absurd conclusion, that it will cover the same distance in the same or a shorter time, which is absurd. For if it covered the distance F [which was] full of substance in the time G, it is surely obvious that if the same distance were void it would have covered it in a shorter time (for that follows); but it was assumed that it had covered the distance F when void in the time G; it will therefore cover the same in a longer and a shorter time, which is absurd.

216a4 It is obvious therefore that if there will be some time in which it will be carried over some part of the void, [this impossibility will follow: in an equal time the result[146] will be got that it will pass through something which was full and void; for there will be some substance proportionate to another as is the time to the time.]

It is clear, he says, that 'this impossibility' followed the hypothesis which said that something moved through a void, that in an equal time the full and the void would be traversed. And taking up again this argument he sets it out shortly: 'for there will be some substance proportionate to another as is the time to the time', that is, it is possible to obtain some proportion between substances which is like the proportion which a time has to a time. So that if I take a substance with fine parts with this ratio to air that the time has in which something moved through the air to the time in which it moved through the void, the things said will follow.

> **216a8** To sum up, it is clear what the explanation of the conclusion is, [that there is a ratio of all movement to movement (for it is in time, and there is a ratio of all time to time, with both being determinate) but there is none between void and full. These things result from the differences between the media in which they move.]

He adds the explanation from which the absurdities mentioned followed from the hypothesis of the void. For because of this there followed the hypothesis of the void the absurdity that if we grant that 660,1
movement occurs in a void, we also grant at once that the void also has a ratio to the full; for it is necessary that if every movement has a ratio to movement, because all movement is in time, and time [has a ratio] to time, consequently the void also has a ratio to the full. Since therefore there is no ratio between the void and substance, and we postulate the non-existent, the absurdity with reason follows from a false hypothesis.

> **216a12** And in connexion with the difference[147] [in speed] between things being carried are the following.

Having said[148] that movement with unequal speeds results from two causes, 'either by differences in the medium, or by differences in the thing moving, if other things are the same', having shown therefore from the things through which movement occurs that it is impossible for movement to occur if void exists, he wants now to come to the rest

and show from the moving things themselves that if void existed it was impossible for there to be motion. For we, who do not say that motion occurs through a void, are able to give very natural explanations for the unequal speeds of the things moving, from the differences in the mediums, as in what we said before, but also from the very differences among the things that move. For if that through which the movement occurs is the same, but the things that move are different either in their weight or in their shape, the heavier thing is carried down more quickly, provided that they are the same in their shapes. And the reason is clear: for since movement is through substance, and the heavier divides more by its power, it is reasonable that it is carried down actually faster. And if they have the same weight, but differ from one another in their shapes, and the one is either spherical or pointed, and the other flat and like a leaf, and if they have the same weight, the pointed and spherical fall down more quickly than the flat; for the flat is held up by a wider substance, and the wider is pulled down[149] more slowly or divided [more slowly] by the bluntness of the shape, but the sharp and the spherical are held up by a smaller substance, so that the underlying [mediums] are divided more quickly. Even those who practise the mechanical arts understand these things; at any rate sailors are equipped with heavy leaden weights,[150] because they are carried down more quickly, and
661,1 fishermen weave lead into their nets. We on the one hand, he says, give this explanation for unequal speeds of movement, but those who say on the other that the movement occurs through a void, what explanation for unequal speeds of movement would they give? For what extra will lead moving through a void have than cork? For the void yields in the same way to both, or rather it does not yield (for yielding belongs to substance), but both in the same way pass through it not being hindered by anything. So all will be moving at the same speed through the void. But that is impossible; movement would not therefore occur in a void.

Those who introduced the void, he says, thought they could prove from motion in place that the void exists, but argument has shown the opposite, that if void exists, it is impossible for movement to exist. In this way he showed from motion in place that the void does not exist, and to those who look into it in itself, he says,[151] the void would have been shown as truly void, that is vain[152] and making no contribution to the things that exist. For since those who introduced the void and [introduced it] as providing the function of place to things that exist, set it out like this, he shows that neither in this way is it possible for the void to exist. He sets out an example like this: if he says, a cube of wood were put in water, as much water would give way to the cube as is the [amount of the] cube itself. Likewise if in air, and if in any yielding substance whatsoever. When therefore the cube is put in the void, will an amount of void equal to the cube give way to the cube? And how would the void give way? If it does not give way when the cube is put in, but stays the same, clearly the cube will occupy the same amount[153] as it does itself; so that the amount of the cube will pass through an equal amount of the void, and the amounts will fit one another. Just as, he says, if the water or the air does not give way to the cube, when it is placed in it, it would pass through the same amount of water or air as it has itself, thus if there is nothing in the void which gives way, the cube will occupy an equal amount to itself. But an amount, he says, will differ in no way from another amount in this actual respect; for even if the amount of the cube has qualities that are adventitious to it, but not as it was originally made, by this it is in place, but only insofar as it has dimensions;[154] so that even if it were separated from all its properties, whiteness, heat, resistance, weight and the rest, and the amount alone remains, nonetheless it is in place and occupies a part of the void equal to itself; for it was in place through none of these, not as white nor as hot nor as having weight, but just as having dimensions.

Since therefore this thing is in place if it is separated from all the other things which make no contribution to its being in place, in what way will place differ from the amount of the void which has come to be in it? If it does not differ at all (for both that amount and this one are without properties) what need was there to surround bodies with another dimension[155] from outside? For if the dimension holds each of the bodies from within, what need was there for another dimension from outside? The external dimension is therefore useless and redundant, since each thing carries round its dimension in itself. And if two similar amounts passed through one another and adapted to one another, why not three as well and an infinite number? And if the void is an amount and differs in no way from the amount that exists in it, void has therefore come to be in void. Again, if the cube changes its place for no other reason than that it is an amount, since just as being an amount it changes place from place, and the bodies in motion alter their own amount together, but this amount in itself differs in no way from the amount as void, to think of another amount from outside for the bodies is therefore redundant and useless, with each carrying along its own amount. For if place is an amount, but the amount of each is in each, with nothing different in this from the outside amount, it is useless to wheel in the other amount from outside alongside the one that exists in each one. For what advantage is there in the like coming to be in the like? And if bodies, as being amounts, need other amounts, the void too, as an amount, would need another; so that the void would come about in a void, and that to infinity. So it is not possible for the void to exist even as place. If therefore it neither provides a function for place, nor is a cause of motion, but on the contrary, if it exists it does away with motion, it seems that the void does not exist in any way.

> **216a13** For we see what has[156] the greater force of weight or lightness, if the others are alike in their shapes, [being carried more quickly over an equal space, and in the ratio which their sizes have to one another. So that [this situation would apply] also in the void. But that is impossible; for for what reason will it be carried more quickly? For in full things of necessity; for the larger divides them by its force ...]

If that through which they are carried is the same, and the things carried are of similar shapes, but unequal in their weights, it is necessary that the one movement should be in the same proportion to the other movement as the one weight is to the other; for by how much the heavier exceeds [the lighter], by that much it will be carried down more quickly. But if they differ in their shapes, it is possible for
663,1 a thing lighter in weight but flat in shape to be carried down through the same distance in an equal time as a larger spherical weight; and it is likely that it would be in an even longer time through the proportion of both the excess of the weight and that of the shape.

> **216a19** For it divides either by shape or by a force [which what is carried or shot out has. All things therefore[157] will have the same speed. But that is impossible. That therefore if there is a void there follows the opposite from what those who said that there was a void argued, is clear from what has been said. Some therefore think that the void exists, if there is going to be motion in place, separate by itself; but this is the same as to say that place is something apart; but it has been said before[158] that this is impossible.]

And the sphere starts its cutting at a point. For this reason both bronze and lead when flat can ride on water, but if they are spherical or cylindrical not so, even if they are much lighter in weight. If therefore that through which the movement occurs were void, clearly all things would have been carried down at the same rate. For what would it have been that would have prevented them moving at the same rate if there were nothing hindering the division? But it is impossible for everything to move at the same rate: so that it is also impossible for there to be a void.

> **216a26** To those who look at the thing in itself it would appear that the void people speak about is truly void. [For just as if someone put a cube in water, the same amount of water will overflow as is that of the cube.]

In itself, that is, apart from motion in place; for above, from the fact that movement in place cannot occur through it, he did away with the existence of void, but now he looks at it like this, as if there were no motion in place. Or since he said that those who say that the void exists because motion exists suppose that the void is 'separate' and existing by itself, but those who say this differ in no way from those who say that place is extension, so that as it has been shown that place is not some extension separate by itself, it would have been shown as well that there is also not a void, for this reason he added that also 'in itself' apart from its having been done away with along with place with extensions, it would have been shown that it cannot exist.

216a29 So also in air; but that is not obvious to sense.

Also in the air, he says, if a substance is put in, the same thing happens; for as much air gives way to the body put into it, as is [the
664,1 amount of] that body, but since our sense is not aware of the air that gives way because it is colourless, to this it appears that nothing gives way. But even if this is not obvious to sense, still it is very obvious to reason; for if air is a substance, and it is impossible for two substances to be in the same [place], there is then every necessity for the air to give way to the substance put in it.

> **216a29** And always in every substance having a capacity for change, it is necessary for it to move[159] towards its natural goal, if it is not compressed, [either always downwards, if its trend is downwards, like earth, or upwards, if fire, or in both directions, whatever may be the thing put in it.]

He said 'having a capacity for change' because of the mathematical quantities;[160] for these clearly do not have a capacity for change, since they [do not have] weight or lightness; they come from abstraction and in thought. In every substance, therefore, he says, in which the cube might be put, the substance gives way to the cube in that direction towards which it was natural to move, for instance if it were put in water, the water would give way to the cube downwards (for water has a tendency to carry a thing downwards) but if a cube of stone were put into fire, the fire would give way upwards and make room for the stone; for it tends to go upwards, unless indeed it is forced; for if the fire had nowhere to go to upwards because a more solid body prevented its upward progress, clearly it goes downwards giving space to the thing put in it. Likewise too in the water in the pot, the water will not give way downwards (for it is prevented by the body which holds it in, which is more solid), but upwards because of the force. The words 'if it is not compressed', [are] as with the wool; for this, when it is compressed into itself, makes way for the thing put in, but does not however move downwards or upwards, but is only compressed into itself. And if air underwent the same?[161] For it is possible that, when the cube is put in, the air goes neither upwards nor downwards, but becomes denser in itself when being compressed, clearly not like the wool, but being changed by becoming denser. But the words 'or in both directions' mean that if that in which it is put is neither too heavy nor too light, but as it were in between in a way and easily able to go in either direction, like the air; for if the cube were put in this, it is possible for the air to be squeezed both upwards and downwards, wherever it would more easily find its way out. The words 'whatever may be the things put in it' if the text is actually like this,[162] he would have used about the shape of the thing put in, that there will be the same outcome, even if what was put in were not a cube, but whatever other shape. But if the words are 'to wherever it may be, that is, to wherever the thing put in has its movement, he would be saying that the substance in which the cube has been put will go towards that towards which the thing put in also goes, clearly because the force of the thing put in is more effective; for if the power of the thing put in were not more effective, that in which it is put would not make way to it. For if the thing put in had a tendency to go upwards, that too will push upwards; likewise too if it had a tendency to go downwards, downwards. The words 'or in both directions' will not fit in very well on this explanation. Or 'in both' was spoken as in a hypothesis that if it were like this, so as to be able to go up and down. 'To move towards its natural goal' is equivalent to 'to move'. 'Having a capacity for change' that is, 'being naturally able to move'; for if it were immobile, the thing put in will not at all get to be in its place.

> **216a33** In the void this is impossible. [(For there is no substance), but an equal amount [of void] must have passed through the cube as was earlier in the void, just as if the water did not give way to the wooden cube, nor the air, but went entirely through it. But the cube also has the same size as the amount of void it occupies.]

Clearly the giving way to the body put in it; for the void is not some body, so that it can give way to the cube put into it. It is necessary therefore that the void does not give way to the cube that has been put into it, but stays put; for how would the void be moved? If it stays put, it is clear that an equal amount of the void as is that of the cube went through the whole cube, or the amount of the cube through an equal amount of the void; for it is the same thing.

> **216b4** If this is also hot and[163] cold or heavy or light, it is none the less different in being from all of the qualities. [Even if not separable; I mean the volume of the wooden cube. So that even if it were separated from all the others and were neither heavy nor light, it will occupy the same amount of void and will be in the same part of place and the same part of the void equal to itself. What difference therefore will there be between the body[164] of the cube and the equivalent void and place? And if there are two like this, why will there not be however many in the same place? This is one thing absurd and impossible.]

Since he said that just as, if the water did not give way to the cube, the two bodies would pass through one another, so since the void does not give way, clearly the amount of the cube and that of the void will
666,1 pass through one another, lest anyone should say 'But it is impossible for a body to pass through a body, the void however is not a body (for it cannot be affected) but an extension; so that it is not the case that if it is impossible for a body to pass through a body, for this reason it is immediately also impossible for an extension to pass through an extension', he says that in no other way are bodies in a place than as they are extensions. For if it is a feature of what is in place to have weight or lightness or any of the other qualities yet it is not as having qualities that it is in place, but as it has extension. For weight or lightness carry a thing into its own place, and being in place depends on nothing else except volume. And we say that it is impossible for bodies at least to pass through one another, not because they have qualities, but because they have extension. The 'impossible' here, therefore, is for two extensions to pass through one another. Since therefore it is one thing for the cube to have extension and another to have qualities, and it is in place not as having qualities but as having volume, if it were stripped of all its qualities, it would in no way differ from the extension of the void; so that if two similar extensions passed through one another, it is clearly also possible for

even more to pass; but that is impossible; for if it is possible for an extension to pass in an extension, it is also possible for a body to pass in a body; for in no other way are bodies in a place than by their volumes; if therefore it is possible for a volume to be in a volume, its other features in no way prevent the body from being in a body. But that is impossible; and so neither is it possible for a body to be in a void; for both are volumes and differ only in having qualities or not. So that if the void exists[165] in order that it may receive bodies, and motion may occur through it, and it is impossible for these to happen in the void, the void then would be useless.

> **216b12** Again it is obvious that the cube, even when displaced, will have what all the other bodies have, [so that if it differs in no way from place, why should one create a place for bodies alongside the volume of each, if the volume is unaffected. For it does nothing if there were another equal extension of its kind round it.[166] That therefore there is no separate void is clear from these [arguments].]

Another argument. Each of the bodies, he says, when it is moving and changing its place, takes with it also its own extension; if therefore the extension in each, separated in thought from its attributes, is different in no way from the void, and the bodies when moving move with their own extensions, what need do the bodies have for other extensions like these from outside? For if the place is the extension, and each has in itself its own extension, it is pointless to think about another similar one from outside; for it contributes nothing to it. And that it is impossible we have already said. For if, because it is an extension each one needs another extension, and the void is also an extension, it too will need another extension, and from this to infinity. And the void will also be in a void.

Chapter 9

> **216b22** There are some who think because of rare and dense it is obvious that there is a void. [For if there is no rare and dense, it is not possible to come together and be compressed.]

Having shown that there is no separate void, he now wants to show also that the scattered void does not exist. Since those who say that the scattered void exists from the rarefaction and condensation of bodies try to show this, he first sets down the arguments by which they tried to establish that there is void scattered about in bodies, and secondly, before refuting their arguments, he shows that it is impossible for the scattered void to exist, and thirdly he shows that even if there is no void it is possible for rarefaction and condensation to exist, from which he refutes also the arguments through which they argued that the void exists. The arguments therefore, through which they argued that the void exists are like this: if the void does not exist, he says, rarefaction and condensation do not exist, but rarefaction and condensation do exist: the void therefore exists. He argues for the additional assumption, I mean that rarefaction and condensation exist, like this: for if, he says, rarefaction and condensation do not exist, neither can bodies come together and be compressed into themselves (for condensation is nothing other than the compression into themselves of the bodies), but if there is not compression, there is every necessity either that there is not motion at all, which is against what is obvious, or if there were motion, it is necessary for the all to swell. And again, in addition to these, either there will not be any changing of the elements[167] into one another, or, if there will be, either the changing of the elements will be into equal volumes, like the cupful[168] of air changing into a cupful of water, and the water likewise into an equal amount of air, or at least, however much water here has changed into air, it is necessary for that amount of air elsewhere to change into water. Aristotle only argues for the additional assumption of the logical argument, because the consequence is obvious, but so that we may make the argument clearer, let us tell it from the beginning like this: if there is rarefaction and condensation, there is void, but the first, the second
668,1 also therefore. How is it that if there is rarefaction and condensation, there is a void? Because condensation is a compression and a shrinkage of the bodies into themselves, and the things that are being compressed and shrinking into themselves are shrunk into the voids scattered about in them, as with a sponge and wool, and things like that; for if the parts of the things being compressed do not shrink into

some voids, they will pass into themselves, but it is impossible for a body to pass through a body. We will say the same things about rarefaction too; for what is being rarefied and extended into a greater volume, if it were not that it became more rarefied by taking in some of the voids in itself, it would follow that there will be a body in a body before it is rarefied; for the parts which have now occupied another place, if they were not in a void before this, they surely passed through one another. But the place which they occupied must have been void before this.

Aristotle establishes the conditional from this thought, and establishes the additional assumption, I mean 'But there are rarefaction and condensation', like this: If there is no rarefaction or condensation, there is also no compression of bodies, and if there is no compression, either movement in a straight line would be destroyed completely – for the things moving in a straight line are moved in this way by pushing the air in front, or the water; so the substance through which the movement occurs is pushed and contracted into the voids lying beside it and gives room for the things going through it, as is also the case with people walking through a crowd – if therefore the bodies are not condensed or compressed into the scattered voids, either movement in a straight line will be entirely destroyed (for where will it go to if there is no room, and with the thing being pushed not having anywhere to go to that would provide an unhindered passage to the moving thing?) or if there is motion, it is necessary, they say, for the all to swell, that is, to get outside of and fall out beyond its own boundaries; for it is necessary, if something is moving, for the body in front to be pushed, and for that in turn to push, and so on as far as the end, and in this way finally for the heaven as it were to overflow into the outside and as it were to move up and down, which also happens in the swimming pools, whenever any body moves in them. They said these things because they did not yet understand the exchanging of places of bodies.

Not only do these absurdities follow the destruction of the void, I mean the destruction of rarefaction and condensation, and that through this destruction either for there to be no movement in a straight line at all, or for the all to swell, but again if the void is done away with, either there is no change at all of the bodies into one another, or, if this clearly does happen, either the change of the bodies comes about into equal volumes, I mean so that the cupful of air, if it changed into water, would produce a cupful of water, and similarly the water would produce an equal amount of air (for where will the larger body go if there is no void?); or if this is against what is obvious (for the change of bodies to something larger is seen; for how are the jars and the wineskins broken by the new wine, unless it is by that wine turning into vapour, with its being larger than what has been changed, the vessels do not have room and are broken, which also happens with things that are touched?).[169] But if the change is into something larger, again one of two absurdities will not be escaped; for either whatever amount of water here changes into air, so much air must turn into water somewhere else (which is itself fanciful, to say that since so much water now appears to me to steam off when boiling, it is necessary for somewhere else the same amount of air to change [into water], and if it seems to me that more [water steams off], for more air also to change [into water]) or again if not this, it is necessary for the all to swell, which itself is laughable in itself, that since this thing has moved, the heaven should be put out of its own places, and that in escaping the smaller void they will accept that they must agree of necessity to the greater. For into what does the heaven swell? For it is impossible [for it to swell] into a body; it is necessary therefore [for it to swell] into a continuous void surrounding the heaven from outside. If therefore all these things are absurd, it is certainly necessary for the scattered void to be in the bodies.

The arguments by which they established that the void exists are like this, and next he shows that the scattered void is impossible. For it is necessary, he says, either for it to be in large parts in the bodies, so that it can make room for another body in it, or very scattered in small parts, so that it is impossible for there to be an entry for a body to slip into it, something such as those who posited pores.[170] If therefore the void were without parts, this would differ in no way from a separate void; we will do away with this too with the same arguments as we did for that. For their dividing the void in this way will make no difference for escaping those problems into which fell those who posited a continuous void; for they too make it a separate extension. If they were to say that the void is scattered in small pieces in bodies, as someone might say, hidden in and mingled in the more 670,1 rarefied bodies, and as it were being a quality of them, those who say this say something more persuasive, though not indeed true. Since, he says, those postulating the void postulate it as a cause of motion, first you must know that it is impossible to postulate it as a cause of motion otherwise than as efficient; for it would not be a cause of motion as place or as that through which; for it has no room in itself for any body, but the causes of motion were like that, either as final, I mean the natural places, or as that through which is the motion. Only therefore as efficient would it be a cause of motion. At least those who posit this say as follows: for they say that light bodies are such because of their entanglement with the void, and the more numerous the scattered voids are, so much the lighter are the bodies. First, he says, if they say that this is the efficient cause of lightness, what will they say is the cause of heaviness? For just as the void interspersed in the light bodies, as in fire, makes them light, so there ought also to be something else scattered in the heavy ones, that makes them heavy. Next, if they say that the void is light, and by its being carried upwards in this way it carries the bodies in which it has been interspersed, many absurdities will follow: for first, what motion of the void could be conceived? Next, it will follow that the void exists in a void. And how can there be a place of the void, or [how can] a void exist in a void? Next, if by how much a thing is more rarefied, and, to say the same, by how much the more they have interspersed voids, by so much is it the lighter, with the uplifting cause being greater, I mean, the void, it is clear, he says, that if the void were a whole, they would be carried most quickly. But it is impossible, he says, for the void to be moved; for it will be demonstrated by the same argument as we also used earlier. For if the void were to move, it would have some ratio to the full, and this is impossible: it is impossible then for the void to move. But Aristotle in this way refutes those who say that the void is scattered in the bodies because those who say this suppose that it is a cause of motion, but if anyone were to say that because of rarefaction the void is interspersed in the bodies, but that it is not however a cause of motion, what will we say? I say what Aristotle himself has said many

times, that first the continuity of bodies will be done away with, co-perception[171] will be done away with, nourishment and increase
671,1 will be done away with (for it would be a filling of the voids, not nourishment and increase), and the other things that he himself has said in his *On Coming to Be*[172] against this view.

> **216b24** But if this [is so], either there will be no motion at all, or the whole will swell, as Xuthos[173] said.

'If this' that is, if there is no compression of bodies and no aggregation into themselves. Xuthos was some sophist.

> **216b26** Or for air and water continually to change into an equal amount, and I mean such as, if air has come from a spoonful of water, at the same time the same amount of water has come from an equal amount of air.

To say 'for air and water continually to change into an equal amount' would appear to be for the air and the water to change into equal volumes, the spoonful of air to become a spoonful of water, but it does not appear to me to mean this, but that it changes into an equal amount of water as was what was changing into air,[174] so that there should come about an equalisation of the volumes and a giving back. And he made clear that this is what he meant by what he added; 'as if air has come from a spoonful of water' and did not say; 'the same amount of air has come', there ought, he says, 'at the same time from an equal amount of air' as was what was changing from the water, to be water coming about, since, he says, if this were not the case, but the water only turned into air, since it openly changes into a larger substance there is every necessity for there to be a void into which passed the substance coming to be. And otherwise, if he did not mean this, but that the change of the substances would be into equal volumes, as it also seemed to Themistius,[175] it was enough to say that so much of air would come from a spoonful of water, and not add that at the same time the same amount of water would also come from air. For if the change comes about into equal volumes, and if the whole of the water changed into air, nothing impossible followed with the air occupying the same place as had the water, but now saying 'if air has come from a spoonful of water', without adding that it is the same 672,1 amount as the water (although he ought to have added that, if he intended to mean this) then he also added that at the same time the same amount of water has come from an equal amount of air, or 'for the void to exist of necessity'.[176] And yet if they change into equal volumes, even if the water alone were to change, none of the other things will follow, nor is it necessary for the void to exist, as I said; for the air will occupy the same place as the water did before it was changed. That he means this, I think, is clear from the very words of Aristotle, but he brought this in not as following on the doing away with rarefaction (what follows the doing away with rarefaction is that either there is no motion or that the all swells) but either as simply following on the doing away with the void, or so that the argument may proceed from a division. Such as either there are rarefaction and condensation, or there are not; if therefore there are not, the things said will follow, and if there are, either there will be a void into which the things compressed are contracted and by which the things rarefied are separated, or rather into which they might also move from outside, or it is necessary that at the same time both the water should change into air, and the same amount of air should change into water, and this seems to be fanciful.

> **216b30** If therefore they say that the rare is what has several separate voids [it is clear that if it is not possible for a void to exist separately as also not a place with a dimension of its own, nor can it be rare in this sense.]

From there he shows that the scattered void is impossible, and he says that if they postulate that the void has been scattered in the bodies in this way, that there are extensions in the rare body which are so many voids separated from body, so that something else can move into them, such as are the pores in sponges, if they are actually full of air,[177] the same arguments, he says, will also fit with these that are also against those who think that place is extension, and against those supposing that the void is continuous and separate.

216b33 But if it is not separate, but there is however some void in [bodies], it is less impossible, [but it follows first that the void is not the cause of all motion, but of that upwards (for the rare is light, for which reason they say that water too is rare) then the void is not a cause of motion as that in which ...]

That is, if it were not separated off, but as it were broken up into such
673,1 fine parts for it not to be possible for an entry for another body to exist in it, for instance if you were to think of some spherical atoms touching one another. For it is clear from the fact that they do not touch one another on their whole periphery, but at a point, that there are some fine voids in between, but it is impossible for another atom to come to be in them; for there will not be room because of their fineness. But this hypothesis, he says, even if it actually seems less impossible, yet still this too will be refuted as being impossible.

217a2 But just as the wineskins by being carried upwards themselves carry what is attached to them, so the void carries things upwards.[178]

It is not in this way, he says, that the void of this kind is a cause of motion, as the extension in which it occurs is said to be a cause, but just as the wineskins which have some weight hanging on them beneath the surface in water, or actually lying on them, by being themselves carried to the surface of the water carry that too, only like this would this kind of void also be a cause of motion for rare bodies; for by being carried upwards it would carry those things too upwards. In this way it is clear that it would be a cause as efficient.

217a3 And yet how is it possible for void to have lift or place? For then a void comes from a void, into which something is carried.

For it is necessary, if the void moves, for it to exchange place with place, so that it will come to be in another void. And what is more laughable than this? And wanting to make the argument more complicated he did not say ‘for there becomes a place of void to which[179] it is carried', but 'a void comes from a void'; for those who speak like this suppose that place and void are the same.

217a5 Again how will they account for its being carried downwards in the case of the heavy?

For if the void is the cause of the upward lift for light things, what will be the cause of the downward movement for heavy ones? For if they were to say that the nature of the heavy bodies is sufficient for their being carried downwards, why is it not that in light things their nature was sufficient for their upward lift and there was no need for 674,1
the void? Aristotle refuted this argument adequately in other works.[180]

217a6 And it is clear that if by how much a thing were rarer and more full of void it will be carried upwards [if it were completely void, it would be carried most quickly. But perhaps it is impossible for this to be moved.][181]

Another argument. If what has more voids moves upwards more, the whole void would have moved most quickly. But it is impossible for the void to move.

217a8 The same argument, as it is that in the void everything is immovable, so is it that the void also is immovable; for these[182] cannot be compared.

With the argument, he says, by which we showed that it is impossible for there to be motion in a void (because the void did not have a ratio to the full, and it is necessary for there to be the same ratio between the times and the movements as there was for the magnitudes; it followed then that there would be instantaneous movement in a void, and this was the same as not moving), with the same argument therefore it will have been shown that it is not possible for the void to be moved. For what is entirely void will have no ratio to what has some scattered voids; so that in the time in which that moves the total void will have been moved instantaneously. But there is no instantaneous motion; it is impossible then for the void to be moved, and it will make no difference whether it is separate or scattered to its not being by nature mobile at all.

> **217a11** [Since we say there is no void] and there have been satisfactory[183] questions about the other things [that either there is no motion, unless there is condensation and rarefaction, or that the heaven swells …]

If we have done away with the existence of the void, he says, but the problems still have in a way what is plausible and reasonable, it is not justifiable to overlook these either. For this reason at any rate he takes them up, so that he may immediately bring in the solution also.

> **217a13** Or there will always be an equal amount of water from air and air from water; for it is clear that more air comes from water. [It is therefore necessary, if there is no compression, to make what occupies the outside when pushed out to swell, or in another way somewhere for an equal amount of water to change out of air, so that the total volume of the whole should be equal, or for nothing to move. For whenever there is exchange of position, this will follow …]

And from there what has been said by me before[184] is clear, that 675,1 bodies do not change into equal volumes, but an exchange occurs; for he gives the reason for this and says that it is clear that more air comes from water; it is necessary therefore for the exchange of air turning into water to come about at the same time, if there is no void. Again, going on he made this very thing clearer by saying 'or in another way somewhere for an equal amount of water to change out of air, so that the total volume of the whole should be equal'.

> **217a19** But the motion is not always in a circle, but also in a straight line.

For if all things moved in a circle, it would not have been necessary for any of these things which have been recounted to occur (for what moves in a circle does not change its place), but since there is also motion in a straight line, there is every necessity either for there to be a void or for something of the absurdities we have recounted to come about.

(At this point comes Philoponus' *Corollary on Void*, which goes on to 695,8.)

> **217a20** [Some for these reasons would say that there is some void,] but we say, from our assumptions, that there is one matter for opposites, hot and cold and the other natural oppositions, [and it passes from what is potential to what is actual, and the matter is not separable, but its existence is different, and it is one in number for colour and hot and cold.] The matter of a body both large and small is the same. [It is obvious; for when air comes from water, the same matter, while not taking in anything, becomes different, but what was potential became actual, and again water from air likewise, sometimes to large from small, and sometimes to small from large.]

Having shown that neither can the scattered void exist, since the problems through which they argued that the void is like this from the rarefaction and condensation of bodies appear to have something plausible, he wants also to solve them and shows that condensation and rarefaction do not come about through an entanglement of voids. This is in addition to the demonstrations in his first argument. For it has been stated, he says, that one and the same matter underlies all the opposites (like hot and cold, dry and wet, soft and hard, and absolutely all the opposites), and the same [matter] is potentially the opposites, being different as a substance from each of them, receiving however each in turn. Just as, therefore, there is one and the same matter for the hot and the cold and for all the other opposites, so, he says, is it the same matter for a body both large and small; for not only is there the same matter for opposites, but also for what is different. Just as therefore in changing from hot to cold, he says, it does not throw off any of the parts in it which are hot, or take in in addition from outside others that are cold, so it does [not][185] change from hot to cold by some inclusion of the things falling on it from outside, but staying the same as a whole through a whole it changes from the hot to the cold and back from the cold to the hot (for neither is there any part of the fire which was not hot, but as a whole through a whole it is hot; therefore in the change to the cold in all its parts alike as a whole through a whole it changes, and the potential of its matter is brought to actual. Similarly also with white and black and the rest; for it is not that the white when becoming black has thrown away parts of the white by which it was white, and added on black parts from outside, but as a whole through a whole has turned in all its parts alike. It is the same also when from less hot it becomes more hot, not throwing away some cold parts and taking on other hot ones, and so increases its heat, but with a backwards turn of its matter and a change from the potential to the actual), as therefore it is with these, so it is with rarefaction and condensation. For the same matter which is potentially the opposites is sometimes rarefied and sometimes condensed; for when it changes from water into air, it is rarefied not by taking in some voids from outside in addition, but changing in all its parts alike as a whole through a whole into a greater bulk. For as a whole through a whole the water has been rarefied; for it is not that some parts of it are air and some water, but the whole of the water evenly became air. So that rarefaction too occurred through the whole of the substrate. Similarly condensation comes about not by the loss of voids, but by a change in its substance; for here too the whole air has been condensed and changed into water in all its parts alike, and it is not possible to find any part that has not changed, and yet if condensation were a collapsing into the voids, the change would not be alike in all its parts.

Then he goes on to a clearer example. For just as, he says, it is possible to make the circumference of the larger circle [that] of a lesser circle (for it is possible to make the semicircle [of a larger] two-thirds of a smaller circle), just as therefore with these although the arcs of the circle have all their parts alike, when the circumference of the larger circle is contracted and becomes the circumference of the smaller circle, with the circumference coming together all alike from all sides, the parts become more curved than before, and certainly there were no [parts] in the larger circle that were not curved but straight, and they were not bent in the contraction of the circumference, but with the whole circumference of the larger circle being evenly curved, the same again becomes evenly more curved, and again the circumference of the smaller circle would be in the same ratio to that of the larger, just as therefore in these the curve is contracted or extended with there being no gap in the circle, this is so also with rarefaction and condensation, and smallness and largeness. For it is not by adding something that it contracts,[186] but nor is it by the expansion of its parts and the entanglement of voids, as occurs with the expansion of the wool, that it becomes larger in this way from smaller, or with the parts collapsing into the void [coming] back to small from large, but one and the same matter for both being potentially either, becomes actually so, as I said also happens with the circumference of the circle becoming less or more curved. Just as often some qualities go with others, like condensation with freezing or whiteness with dampness or dryness with heat, so also, he says, lightness goes with rarefaction and weight with condensation. These, I mean weight and lightness, always go, the latter with rarefaction and the former with condensation, and two other things go with these for the most part, with condensation the hard, and with rarefaction the soft. But not with all, he says, for there is a difference with iron and lead, for lead, although dense, is soft, but iron is rarer than lead, but harder. So that two qualities go with rarefaction and condensation, with rarefaction always the light, and in general the soft, and with condensation always the heavy and in general the hard. These things Aristotle said fittingly; for since those people said that the rare was light for this reason, that it has more scattered voids (for these are the causes of the upwards lift), for this reason, he himself says, that lightness goes with rarefaction, and weight with condensation, and of these lightness is a cause of the upwards lift, and weight of that downwards. So that, he says, it has been proved sufficiently that there is no void either separate or scattered in bodies, but if, he says, it is dear to someone just as the cause of motion to call that void, let that man call weight and lightness the void, or rather the matter receptive of these. Lightness, he says, and weight are causes of motion in bodies, and hardness and softness of being affected and not being affected; for soft things are more easily affected, and hard less easily.

Aristotle says these things, and from these the problems would have been solved. And it is obvious that there will be motion even if there is no void (for air and water change places with the thing in motion), but those people did not yet understand the exchange of places by bodies, and wondered how motion could occur if there were no void; for if what is in front is pushed by the thing moving it is necessary for the heaven to swell. If therefore bodies change places with one another, both clearly there will be motion and the heaven does not swell. There remains the [problem] about the bodies which change from a smaller to a larger body, as from water to air. They say, therefore, that since rarefaction and condensation come about 698,1 through a change in bodies, and not by passing into the voids, when air comes from water, the external air is compressed when pushed by [the air] that is coming to be, and is condensed by the pressure, and as it is condensed it occupies a smaller place; and in this way neither are there two substances in the same [place] nor does the all swell, but the increased substance has come to be in the place of the one that has been diminished. This seems to me to involve a considerable problem: for I cannot understand by what reasoning a push condenses. For the push moves the thing moved, but motion naturally heats, and what is heated is rarefied; how then can condensation come from the push? For, on the contrary, from our reasoning it appears to be rather a cause of rarefaction, not of condensation. Again, in summertime when the heat is considerable, and the air has the finest parts, how would it not be laughable to say that the water turned to steam by me, when steam comes about or even air, that it pushes the external air and condenses it? How is it not the opposite, that the steam assimilated to the surrounding air is much more rarefied by the friction of the sunbeams; again, since in the summer water is turned into steam and changes into air, if by this there came about room for the air that changed from water by pushing the external air and condensing it, the air should rather be condensed in the summer. For at that time there is a lot of water changing into air. So that air was going to be denser in summer than in winter. For if this argument is correct, let me take the state of the air in winter, then let the water start in the time of spring and that of summer to change into air; it is surely necessary, by how much the greater is the change, the more should the external air be condensed; so that it will follow that the air in summer will be denser than the air in winter, and that is irrational. It is not the case, then, that when water is changed into air, the substance that results from the push of the external air is condensed, and by this makes room for itself. Either therefore the basic elements change into things of equal volume, or we must think that how much water turns into air here, so much air changes into water somewhere else. But the one is impossible, the other fanciful. For that water when it changes into air changes into a greater volume, perception testifies; for the jars and the wineskins with new wine are broken with the wine turning into vapour and not having space in the skin. And the flame and the smoke are much greater than the volume of the wood and the oil; for it is not, as some say, that the smoke, scattered like a cloud of dust, occupies a larger place than the body that gave it birth, [and that] not because it was greater in volume, but as not being continuous. For this is obviously false. For above all the ratio between the dust gathered together and [the dust] scattered is not as great as that between the smoke and the body that gave it birth. And then we also refute it from our agreed [assumptions]; why are the jars and wineskins filled with the new wine broken if it is not that the vapour coming from the wine was greater in volume than the body that gave it birth? Hence they are in the habit of not filling the jars or the wineskins to the full with the wine, because the vapour that comes off occupies space, and they do not completely seal off the mouths of the jars, but allow a way out for the vapour. Most clear is the turning of nourishment into vapour. For obviously the nourishment changes to a greater volume when turned into vapour. At any rate we observe the volume and the extension of the stomach. That the change of substances comes about not into things of equal volume is clear, but to say that when I have turned a little water or oil here into steam, necessarily somewhere else the same amount of air turns into water, is fanciful. How too is it not irrational that because I turn a little water into steam, the whole of the air is stirred so as to occupy the place of the air changing into water somewhere else? This then for the time was the problem until it met the solution it needed.

We say, from our assumptions, that there is one matter for opposites, that is, those stated in the first book, that there is a single matter underlying the opposites, and that it is potentially being (for form is being) and becomes actually being, as substance different from the things there are, I mean the forms, but never being without a form.[187]

> **217a26** There is the same matter for body, both large and small. [It is obvious; for when air comes from water, it is not that the same matter by taking in something extra becomes different, but what was potential became actual, and again water from air similarly, now to largeness from smallness, and now to smallness from largeness.]

That not only for the opposites is the matter the same, but also for a body. And since they change in volume, sometimes becoming larger and sometimes smaller, and of the large and the small, he says, there
700,1 is the same matter changing to either. There is evidence for this, he says, in that in the changing of the substances we see nothing being added from outside when one gets bigger, as when water becomes air, nor anything being taken away when one gets smaller.

> **217a31** Likewise therefore also if air which is large in volume becomes a smaller volume, [and larger from less, the matter which was potential becomes both. For just as the same [matter] from cold becomes hot and from hot becomes cold, what it was potentially, so also from hot yet hotter, with nothing in the matter becoming hot which was not hot when it was less hot ...]

That not only when the change is complete does this follow, but also when the same things become [something] more and less, as from more rare air comes less rare, and from hotter less hot; and there the matter which was potential changes into either.

217b2 ... just as the circumference of the greater circle and its convexity, [if it comes to be of a smaller circle ...]

The example is very clear; for when the less convex becomes more convex, neither have any curved parts from outside come to be in it, nor was there anything in it which before this was not curved but straight, and later became curved, but even before this it was curved, and afterwards became more curved.

217b4 ... being the same or different, [in nothing has the curved come to exist which was not curved but straight (for it is not by leaving out the more or the less) nor is it possible to get a bit of flame in which there is not also heat and whiteness. So therefore also the earlier heat [is related] to the later. So that both the largeness and the smallness of the perceived bulk were ended without its matter taking on anything, but because the matter is potentially of both; so that the same thing is dense and rare, and they have one matter. The dense is heavy and the rare light.]

That if the same circumference from being less curved becomes more curved, or there arises another circumference that is more curved than the other circumference, the same thing happens. For it is not by there being some gap in the circumferences which was not curved before, and so becomes more curved, but having been evenly less curved before it again changed in all its parts alike afterwards and became more curved; for the less curved is not said to be less curved by its having gaps which are not curved, but by its being bent gently as whole through whole.

217b12 Just as the circumference of the circle when it is con- 701,1
tracted into a smaller one [does not receive any other thing that is hollow, but what was continuous, and whatever of fire anyone might take will all be hot, so also is it that the whole is a contraction and expansion of the same matter.][188]

He says the same thing again in these words, so that the commentators regard this remark as spurious from the part I have appended as far as the 'is a contraction and expansion of the same matter'. Perhaps it is a repetition.

> **217b16** For there are two things on each side, the dense and the rare; [for both the heavy and the hard seem to be dense, and their opposites, both the light and the soft, seem to be rare.]

This immediately follows the above. For after saying 'the dense is heavy, and the rare light' he adds 'for there are two things on each side', that is from each side, the rare and the dense, two qualities arise; for the light and the soft accompany the rare, and the heavy and the soft the dense.

> **217b19** The heavy and the hard are separate in lead and iron.

That is, that the heavier is not always the softer; at any rate lead is heavier than iron but softer.

> **217b20** [From what has been said it is clear that there is no hidden void,] neither absolutely nor in the rare, nor potentially.

'Absolutely', because there is no continuous void outside bodies, in the way that some supposed the air to be, not a kind of substance but a void, or as the Pythagoreans [supposed], what is outside the heaven, but neither in bodies are there any voids with large parts, into which parts of the rare body fall and make the whole more dense. And the 'nor potentially' instead of the 'as mixed in', or that neither as fulfilling the function of place, which is void, but never apart from body; for there was a final discussion about this, where he said that the void will have been shown 'as truly void'.

702,1 **217b21** Unless someone wanted to call the void entirely the cause of motion. [In this way the matter of the heavy and light, as such, would be the void; for the dense and the rare by this opposition are creative of motion, but with regard to the hard and soft [are creative] of being affected and not being affected, and not of motion, but rather of alteration.]

If anyone wanted, he says, to give a new meaning to words and call the void the cause of motion, such a one would be saying that the matter of heavy and light was void; for these are causes of motion. For with regard to the hard and the soft, matter is not a cause of motion, but it is a cause of things being easily affected or affected with difficulty.

> **217b27** And about void, in what way it exists and in what way it does not exist, let it have been worked out in this way.

It has been said that it does not exist as an extension separate from body, either outside bodies or scattered in bodies.

Notes

1. I have chosen 'void' most frequently for *kenon,* though in places 'empty' is better. Sorabji uses 'vacuum', but that risks bringing with it the suggestion of something with a power to suck in.

2. Usually, as here, 'body' is the best word for *sôma,* but at times 'solid' and 'substance' will be used.

3. Of Clazomene. He and Empedocles are said to have attacked the idea of the void, and perhaps it was Melissus whom they were attacking. See 213a22ff.

4. The idea of ordinary men.

5. David Furley (private communication) considered for *diastêma* extent, extension, expanse, volume, interval, distance, and settled for extension. I usually prefer that word, but in places dimension seems better, and sometimes amount or distance.

6. Is this a reference to a children's game? Theophrastus *Char.* 5.5 has the complaisant man playing with children and saying 'wineskins'.

7. Here 'empty' seems preferable for *kenon.*

8. For another account of a clepsydra see Philoponus' *Corollaries* 569,21ff. The function of a clepsydra was to measure time, and for that purpose many versions were developed. Philoponus is only interested in it as evidence for the fact that air is a substance.

9. Normally *apophrassein* means 'to block up', but in this context it seems to be contrasted with *emphrassein,* and since *apo* can have a negative function 'unblock' fits the context and should be preferred. On the other hand, at 612,21 and 25 'stop' is required.

10. The Atomists and the Epicureans.

11. This is the view of the Pythagoreans. See *Phys.* 213a22-6, and below 610,8-22; 622,13.

12. It is not clear how to take this. It seems to be Philoponus' own remark. The passage is from 213b2.

13. 'motion' here is for *kinêsis,* which can also mean more widely 'change'; 'in place' specifies its kind.

14. If the first, the second: but the first, the second therefore.

15. An Eleatic philosopher. His argument is given at length at *GC* 325a2-16.

16. Vitelli says it is rather *Phys.* 214b8.

17. The argument comes from the belief that when water was added to dry ashes the total volume was not increased. See *GC* 325a23-32, b9-10, *Probl.* 938b24-939a9. The argument has been attributed to Leucippus.

18. This is based on *Phys,* 213b22ff. given below.

19. This symbolic treatment of the void is less interesting than most of the reported examples, such as that the sea was the tear of Kronos and the planets the dogs of Persephone in Porphyry *Life of Pythagoras* 41.

20. The grammar of the text as we have it is peculiar, but the sense is not affected.

21. A stock example of an imaginary creature.

22. While clepsydras are famous as water clocks, Aristotle is only concerned here with a part of their mechanism, involving a siphon. Philoponus gives a similar account in his *Corollary on Place* at 569,21-570,9. But 612,15-16 are puzzling and do not fit in with what we otherwise know about clepsydras, and I wonder if they should be rejected.

23. Harpagion: Alex. Aphrod. *Probl.* 1.95.

24. Ross mentions Metrodorus of Chios, a pupil of Democritus.

25. Vitelli follows the MSS evidence and prints *eipe* 'he said', but suggests *eipon* 'I said', which I prefer. The reference is to 610,8.

26. There is little evidence about Zeno himself, but Philoponus is probably referring to the Stoics in general. For them the best evidence is from Cleomedes 6,11 and 10,24-12,5.

27. The majority of the evidence for the text of Aristotle has *dokei pasi* 'appears to all'. That is an excessive claim, and perhaps we should prefer Philoponus' *dunatai* 'can'. On the other hand what follows is in *oratio obliqua* which supports *dokei.*

28. *Anal. Post.* B1.

29. It is well known that Philoponus indulges in puns. The Greek word here is *atopos,* which can be translated as 'placeless', an appropriate word in this context. Aristotle himself used the same apparent pun at *Physics* 214a4. See 620,10 below.

30. It might be assumed from this that the people who believed in the void also said something about the point. But in fact Aristotle (214a4-6) introduced the point as a problem for them, as Philoponus says at 620,12.

31. The text is faulty here. The argument is that to avoid having to say that a point is a void one must define void not just as a place, but as an extension. I suggest reading in 617,11 *topon diastêma en hôi mêden esti bareos kai kouphou* 'place is an extension in which there is nothing heavy or light'.

32. See 621,6.

33. That would exclude what was just coloured or sounding.

34. It is doubtful if these people ever existed. If they did, we certainly do not know who they were. See *Phys.* 214a13-14, and below 621,20-3.

35. 214a15.

36. 211b18.

37. This reference cannot be traced. But see 622,19, where *kata to hupokeimenon* has the same meaning.

38. 211b19ff.

39. This sentence seems wrong. The upshot of this passage at 619,13-14 is that the void is rejected. So 619,10-13 must involve the rejection of arguments in favour of (a) place being extension and (b) the existence of the void. The whole sentence 619,10-13 is omitted by some MSS.

40. These must be Democritus and his followers. Their views were adopted by Epicurus and his followers, with which Philoponus was familiar. Lucretius' *De Rerum Natura* is a valuable secondary source.

41. *De Caelo* 290b12.

42. i.e. having some other body, like one coloured or sounding, but without weight or lightness.

43. The majority of the MSS of Aristotle have *mê tode ti* ('not a this something').

44. This is puzzling: Aristotle himself was the first to speak of matter, but Ross p. 585 explores possible explanations. In what follows Philoponus does not help.

45. Referring to the debatable 'first, or primary, matter' which some attribute to Aristotle.

46. Following Ross's insertion of commas.

47. 211b14-29.

48. 212b6.

49. The Pythagoreans and the Stoics.

50. *paremplokê,* which is Epicurus' own word.

51. *hupokeimenon*, a technical Aristotelian term. Cf. 619,5.

52. See Sedley (1987) for a discussion of this paradox.

53. Simplicius, *Corollary on Place* 618,20-5 attributes this view to some Platonists and Strato, who is also mentioned at 601,24. Philoponus at 579,6ff. using the expression *topikon diastêma* 'place extension', defends the concept of filled extension. Cf. 569,8ff. and 675,22-676,1.

54. Philoponus follows one of two readings at this point, the one that makes the subject of the verb 'to want' or 'prefer', the void. I have preferred the other, in which the subject is those who believe in the void. Neither is easy.

55. This sentence does not hang together well, but I leave it as it is.

56. This may refer to 210a31, where there is a question about whether a thing can be inside itself.

57. The Greek is *eph' hou*. Aristotle has *en hôi* 'in which'. Philoponus has taken over from Aristotle the idea of a cause 'in which', which Aristotle attributes to those who believe in the void. It is not a part of Aristotle's own system. Philoponus seems to need to explain the preposition by another one.

58. *alloiôsis*. This is seen as one kind of *kinêsis* 'motion'.

59. *Phys.* 214a31-2.

60. Aratus' poem *Phaenomena* was a versification of a work by the astronomer Eudoxus. It is quoted at 658,6-8. Much of it is about the circulation of the constellations.

61. That is, the place occupied is the same, but there are movements within it.

62. Lucretius *On the Nature of Things* 1.372-83. The word *kêtos* has had a variety of meanings, but applies to large sea creatures.

63. The MSS all have *autois* 'to them'. Vitelli has suggested as an alternative *autôi* 'to it' which I have adopted.

64. Commentators agree that the example is of a wine-jar which contains when full a quantity of wine, but does not overflow when wineskins which have held wine are added. But they differ over some details.

65. This seems the best word for *ousia* here.

66. The pressure of the wineskins, or, as a reader suggests, 'in this way'.

67. I accept Vitelli's alternative reading *eplêrou.*

68. 126,31.

69. This probably means 'throughout itself'.

70. *alloiousthai.*

71. *alloiôsis.*

72. The word is *kinêsis* which can usually be rendered by 'motion' or 'movement', but in places 'change' can be more appropriate. In what follows, at 629,1 and 3 the word *akinêtos* is found twice. For the first I have used 'motionless', and for the second 'changeless'. Later, at 641,2 I have used 'unmoved'.

73. *alloiousthai.*

74. Melissus and others.

75. This is Aristotle's word, for which I have used 'expel'. Philoponus thinks it needs explanation.

76. The Atomists and the Epicureans. See Lucretius 1.334-90.

77. Philoponus has adopted two of Aristotle's four causes, but then has 'that in which' and 'that through which'. Aristotle himself at 215a29 says that that through which is a cause. See also 670,3-9.

78. This sentence is puzzling. The MS M has an incomprehensible jumble where others have *megethei* (size) and possibly a chunk of text has been lost. Or we might take *megethei* as 'in something extended'.

79. i.e. homogeneous.

80. On the Aristotelian view the limits of bodies are their surfaces, and that means that those limits are intimately related to the bodies they enclose.

81. The MSS differ between *ton* (masculine) and *to* (neuter), which suggests that the meaning was unclear. The thought appears to be that behind the circular movement there is something which brings about that movement. Sorabji, *Matter, Space and Motion*, 212 reads *tên kuklôi* at 632,11 and 12 and translates 'the surface at the circumference'.

82. 212a31.

83. This sentence is difficult. I have repunctuated it.

84. 199b28.

85. 212b30.

86. 214a16.

87. This does not make sense. We need a negative with *edei* 'should', although that is not easy to fit into our text.

88. Ross notes that Simplicius 665,33 takes this as meaning: 'things do not turn out right for those who believe in a separate void'. Hussey has 'It does not work out'.

89. At 636,12 Philoponus paraphrases this passage and has *sôma ti* 'a body' where the lemma has *sômati*. Ross has adopted the former reading, which seems preferable, although difficulties remain. The commentators have problems, and Simplicius 665,26-9 indicates that 214b25-9 are absent from some MSS. The point involved is a distinction between how a whole thing is in place and how a part of that whole is in the whole, but it is not clear how that fits in.

90. 211b19. There are considerable difficulties with this passage. See Ross 572-3.

91. 548,16.

92. Aristotle's text has *kenon* 'void', but Philoponus' lemma omits it.

93. Plato *Phaedo* 109A, *Timaeus* 62D. Aristotle *De Caelo* 295b11 refers to Anaximander.

94. This sentence lacks a verb. It can be seen as an introduction to what comes next.

95. This is unusual. It may derive from *De Generatione Animalium* 742b10, where Aristotle distinguishes between *kinêtika* and *organika*. Simplicius ad loc. 664,1ff. uses the expressions 'in which' and 'through which' as Philoponus does at 630,18.

96. 215a14.

97. The reading accepted by Vitelli is ungrammatical: I prefer to adopt a finite verb.

98. This reference cannot be traced. Very little survives of Philoponus' commentary on the eighth book of the *Physics*, and we do not know what is meant by his word *skhôlais*, for which I have used 'notes'.

99. 254b12.

100. The point is to illustrate the complexity of the supposed movement of the air. The original idea is that of musical instruments, but it was then transferred to the shape of running tracks.

101. Wolff in Sorabji, *Philoponus*, 98 n. 47 asks whether this sentence means that both alternatives involve unnatural motion. He thinks the second does not. In any case that point is not the centre of the argument of the passage.

102. This is puzzling. Sorabji has: 'what has that to do with the thing thrown? For the air there falls short of it.'

103. *DA* 418a34-419a12.

104. Cf. Philop. *On the Soul* 334,40-335,30.

105. This sentence is the last one in a parenthetical section.

106. Aristotle's text is itself uncertain. Philoponus must have read *ta phusei*, and felt it needed explanation. The better reading, *phusei,* would be about motions, not things.

107. Again Aristotle's text is uncertain. *Ta rhiptomena* 'the things thrown' makes the interpretation easier.

108. Ross refers to Plato *Timaeus* 79A-80C.

109. This is obscure. I suggest deleting *ouk* 'not' in 644,6. There is then a contrast between a situation in which there is a distance between the hand and the stone and the actual one in which there is no distance.

110. This terse statement is interpreted by both Philoponus and Simplicius as referring to the view of the Atomists that there are scattered voids. At 644,27 Philoponus claims to be quoting Aristotle's own words, but there is nothing similar in our text at this point. Themistius and Simplicius also have something on these lines, and it is possible that something has been lost from our text of Aristotle. Alternatively it came from Aristotle's lost work on Democritus.

111. In this context *sôma* includes solid bodies but also air and water. See n. 119.

112. That is, movements at different speeds. I have kept 'unequal' throughout, but this is what it means.

113. Cf. Lucretius 1.358-63.

114. This must mean a line made by a point. Cf. 647,4, where the example is of the point at which a tangent touches a curved surface.

115. Here this seems the best way of translating *diastêma.*

116. A length.

117. The word *logos* can mean both relation and ratio. I have tried to use the latter in more mathematical contexts.

118. This is based on Aristotle's long passage starting at 214a30. I have kept the same order in our letters as that in the Greek alphabet, so that for example gamma becomes C.

119. In this section *sôma* is used of water and air, which have low density and no shape, so that 'substance' seems preferable to 'body'.

120. Here and in what follows I again use 'distance' for *diastêma.*

121. 216a5-6.

122. This addition seems required by the sense. It is omitted in MS M.

123. Cf. 215b4ff.

124. 216a8.

125. This may involve another pun.

126. There is nothing like this in our text of Aristotle. His reference to earth at this point is anyhow puzzling, as Philoponus hints. Themistius 130,23 also refers to mud.

127. Vitelli keeps the main MSS reading and notes that Philoponus 'in his way' ignores the structure of the sentence he has started.

128. I think this means 'if you consider air as a substance'.

129. Aristotle's text has D here, and in lines 15 and 16 below we should also read D for C.

130. The Greek MSS of Aristotle all have *kai* ('and'), but Philoponus seems to have read *alla* ('but') as does the Arabo-Latin version collated by Mansion. See Ross 103.

131. What is finer.

132. Here this renders *logos*.

133. 648,15-24.

134. The word is *anomogenê,* and it is contrasted with *homoeidê* 'cognate' in 655,5.

135. In this context 'solid' is preferable to 'body' for *sôma.*

136. The word is *hexis*, which in this context has little more than a positive meaning.

137. I have changed the punctuation in accordance with Ross's edition.

138. See n. 60.

139. Lit. 'with equal days'.

140. *Phaenomena* 225-7. The Little Bear constellation revolves round the North Pole in a small circle, but keeps pace with the Ram (Aries) which moves round a greater circle and so moves faster.

141. In some later passages F is a fine substance. At 215b28-30 it is a *sôma* which is very fine. The important point is that it can be traversed.

142. Aristotle's text is uncertain. Ross, with some MSS, wants to strike out *tôi megethei* 'in size' in 215b23. Philoponus clearly read it, but then exchanges air and water in his interpretation of the B and the D as compared with 215b4.

143. Another pun?

144. I am not sure what this is. At 658,7 he gives what he calls the third argument.

145. Vitelli found this text difficult; there is even an error on what we now have, with *pros* repeated at the end of 657,21 and the beginning of 22. He wanted to delete *atopon kai* 'absurd also' in line 18, and insert the numeral zeta (F) at that place, followed by a closing parenthesis.

146. This is Ross's version of the puzzling *lephthêsetai.*

147. Literally 'excess'.

148. 215a26.

149. The subject should be the flat object. The subject of the following 'divided' should be the medium.

150. As anchors. A fuller account is given by Themistius at *Phys.* 132,17ff.

151. 216a26-7.

152. Aristotle had already joked about the word *kenon* 'void'. Philoponus' joke is different, but on the same lines.

153. In this context 'amount' seems the best word for *diastêma,* which here involves a three-dimensional object.

154. Here and at 662,2 the word related to *diastêma, diastaseis* (pl.) is used.

155. Here *diastêma* brings with it the idea of a box-like container. With some hesitation I use 'dimension'.

156. Philoponus has the singular here, Aristotle's text the plural.

157. This does not follow from what precedes it immediately, which is about movements in spaces that are not void.

158. 211b18-29.

159. The lemma has *sunistanai,* which is supported at 665,13 by Philoponus' own *sunistanai,* but Aristotle's text has *methistanai.*

160. This is Philoponus' explanation for Aristotle's words. Ross, on the other hand, thinks it is meant to exclude celestial matter which 'has no tendency to move up or down'.

161. I have inserted a question mark here.

162. The words are not in the lemma. Further, it appears that Philoponus found two different readings in the MSS available to him, and thought both worthy of comment.

163. Philoponus' MSS have *kai* 'and', but those of Aristotle have *ê* 'or', which is preferable.

164. Here 'body' seems most appropriate.

165. Vitelli has a comma here, but that seems unnecessary.

166. At this point some MSS have several lines which are ignored by Philoponus along with Simplicius and Themistius. I have therefore omitted them.

167. Earth, air, fire and water.

168. A *kyathos* was a small cup which was also used as a measure.

169. Presumably a reference to normal breakages. Can it be a light-hearted gloss?

170. *Poros,* a word used by Epicurus, and presumably by the Atomists.

171. This does not make sense. Vitelli suggested *sunôthêsis,* a very rare word meaning compulsion or forcing, which hardly seems appropriate; *aisthêsis* 'sensation' is a possibility.

172. *GC* 1.5 320b26ff., where Aristotle refers back to this passage in the *Physics.*

173. We know nothing more of Xuthos. Simplicius says he was a Pythagorean, but Philoponus here a sophist. Themistius also discussed him. See n. 175.

174. i.e. the same quantity of water as had before changed into air.

175. 135,24.

176. This reads oddly, for Philoponus has quoted from the lemma without acknowledging that we do not have a simple alternative here.

177. The point of this is unclear, and there is some uncertainty about the text. Can it be a gloss?

178. There is a difference of reading between Aristotle's MSS and what Philoponus has but the sense is the same.

179. Vitelli's text has *on,* but that must be a typographical error for *hon.*

180. *De Caelo* 309a5.

181. Ross, relying on what follows, says this must mean 'it is as impossible for a void to be moved as for things to be moved in a void'.

182. Philoponus has *tauta* 'these', but Aristotle has *ta takhê* 'speeds'.

183. Philoponus has *kalôs* 'well' where Aristotle has *alêthôs* 'accurately'.

184. 671,12.

185. A negative is needed here. I suggest inserting *ou* after *thermou* in 695,27. Further it might be better to have a change from cold to hot to balance that from hot to cold at 695,25.

186. This is odd. Perhaps something has been lost, so that we would get: 'it is not by adding [something that it gets larger, nor by losing something] that it contracts'.

187. This is a summary of views set out in *Phys.* 1.

188. Lines 217b12-16 interrupt Aristotle's argument and appear to repeat 217b2ff. Simplicius 690,30-691,4 thinks it is a marginal note and regards its language as supporting proof of that.

Bibliography

Baltussen, H., *Philosophy and Exegesis in Simplicius,* London 2008.

Bowen A.C. and Todd, A.B., *Cleomedes: Lectures on Astronomy. A Translation of The Heavens with an Introduction and Commentary*, Berkeley 2004.

Charlton, W., *Aristotle's Physics I, II,* Translation with Introduction and Notes, Oxford 1970.

Charlton, W., *Philoponus: On Aristotle On the Soul 2.7-12,* Ancient Commentators on Aristotle, London 2005.

Furley, D.J., *Philoponus: Corollaries on Place and Void,* Ancient Commentators on Aristotle, London 1991.

Golitsis, P., *Les Commentaires de Simplicius et de Jean Philopon à la Physique d'Aristote,* Berlin and New York 2008.

Hussey, E., *Aristotle's Physics Books III and IV,* Oxford 1983.

Kidd, D., *Aratus: Phaenomena,* Cambridge 2004.

Reale, G., *Melisso: Testimonianze e Frammenti,* Florence 1970.

Ross, W.D., *Aristotle's Physics,* Oxford 1936.

Sedley, D.N., 'Philoponus' Conception of Space', in Sorabji 1987.

Sorabji, R., *Time, Creation and the Continuum,* London 1983.

Sorabji, R. (ed.), *Philoponus and the Rejection of Aristotelian Science,* London 1987.

Sorabji, R., *Matter, Space and Motion,* London 1988.

Sorabji, R. (ed.), *Aristotle Transformed,* London 1990.

Stallings, A.E., *Lucretius: On the Nature of Things,* London 2007.

Todd, R.B., *Themistius: On Aristotle Physics 4,* Ancient Commentators on Aristotle, London 2003.

Urmson, J.O., *Simplicius: Corollaries on Place and Time,* Ancient Commentators on Aristotle, London 1992.

Verrycken, K., 'The Development of Philoponus' Thought and its Chronology', in Sorabji 1990.

Wolff, M., 'Philoponus and the Rise of Preclassical Dynamics', in Sorabji 1987.

Bibliography

English-Greek Glossary

above: *anô, anôterô*
above, be (v.): *huperanekhein*
above, have a place (v.): *huperanekhein*
absolutely: *haplôs*
abstraction: *aphairesis*
absurd: *apoklêrôtikos, atopon, geloios*
accidentally: *kata sumbebêkos*
accompany: *huparkhein, parakolouthein, sumparomartein*
account: *logos*
accurately: *akribôs*
accuse: *enkalein*
achieve: *anuein*
activity: *energeia*
actual: *energeiai, kat'energeian*
adapt (v.): *epharmozein*
add: *prostithenai*
added, be: *prosgenesthai*
advance (n.): *epidosis, proodos*
advance (v.): *proienai*
advantage, be: *lusitelein*
adventitious, be: *episumbainein*
affected, being: *pathos*
  not being affected: *apatheia*
  what cannot be affected: *apathês*
  affected with difficulty: *duspathês*
  easily affected: *eupathês*
agent (intellect): *poiêtikos*
agree: *homologein, sunkhôrein*
aim: *katastokhazein*
air: *aêr*
  like air: *aêrôdês*
alike: *homoios*
alike in all its parts: *homoiomerôs*
all, the: *to pan*
all sides, from: *pantakhothen*
allow: *ean*
already exist in: *prouparkhein*
alter together: *summetapherein*
altered, be: *alloiousthai*
amount: *diastêma*
amphora: *amphoreus*
ancient: *palaios*
angleless: *agônios*
  with many angles: *polugônios*
animal: *zôion*
apart: *apolutôs, khôris*
apodeictic: *apodeiktikê*
append: *paratithesthai*
appropriate, be: *harmozein*
aratos: *aratus*
arc: *tmêma*
argue: *epikheirein, kataskeuazein*
argument: *epikheirêma, epikheirêsis, logos*
arise: *sunistasthai*
arrange: *tithesthai*
arrangement: *suntaxis*
arrive: *aphikneisthai*
arrive at: *katantan, tunkhanein*
arrow: *belos*
art: *tekhnê*
ashes: *tephra*
ask: *aporein, zêtein*
assemble: *sunagein*
assign: *apodidonai*
assimilate: *homilein, proskrinein*
association: *sustasis*
assume: *enupotithenai*
assumption: *hupolêpsis*
  additional assumption: *proslêpsis*
at all: *holôs*
atom: *atomos*
attempt: *epikheirêma*
attach: *proskrinein*

attribute: *pathos*
avoid: *apopheugein*
aware of, be: *antilambanein*

back: *empalin*
back (claim): *epagein*
be: *huparkhein*
be in: *eneinai*
bear: *arktos*
become in: *engignesthai*
beginning: *arkhê*
behind: *katopin*
 from behind: *opisthen*
belief: *pistis*
believe: *nomizein, pisteuein, tithesthai*
believe in: *doxazein*
belong: *huparkhein*
bend: *kampein*
beneath the surface: *hupobrukhios*
better than, be: *perieinai*
between, in: *mesos*
black: *melas*
blacken: *melainein*
block: *emphrassein*
blow up: *phusan*
blunt: *amblus*
boat: *ploion*
bodiless: *asômatos*
bodily: *sômatikos*
body: *sôma, ousia*
both: *amphoteros*
bottom, at the: *hupokatôthen*
boundary: *horos*
boundless: *apeiros*
break: *rhêgnunai*
breath: *pneuma*
breathe in: *anapnein*
breathlike: *pneumatôdês*
bring in: *eisagein, epipherein*
briskly: *oxeôs*
bronze: *khalkos*
brought back, be: *apokathistanai*
bubble: *pompholux*

call: *kalein*
carried, be: *pheresthai*
carried on, be: *epokheisthai*
carry: *okhein*
carry along: *perikheein, sumperipherein*
carry down: *katapherein*
carry out: *ekdiaphorein*
case, be the: *sumbainein*
cause: *aitia, aition*
 think a cause: *aitiasthai*
chaff: *karphos*
change (n.): *alloiôsis, metabolê*
change (v.): *metaballein*
 capacity for change: *metastasis*
change places: *antiperiistanai*
check: *kôluein*
circle: *kuklos*
claim: *pistis*
clear: *eudêlos, prophanês*
 very clear: *prodêlos*
 make clear: *saphênizein*
clepsydra: *klepsudra*
coalesce: *sumphuein*
cognate: *homoeidês, sungenês*
cold: *psukhros*
collapsing: *sunizêsis*
collide with: *sumpiptein*
colour (n.): *khrôma*
colour (v.): *khrômatizein, khrônnuein, khrôzein*
come: *epigenesthai, khôrein*
come about: *genesthai*
come in: *eisienai, epeisienai*
come first: *proêgeisthai*
come together: *sunienai*
come upon: *aphikneisthai*
coming to be: *genesis*
commentator: *exêgêtês*
common: *koinos*
compare: *sunkrinein*
comparison: *sunkrisis*
complicated: *plêktikos*
complete: *pantelês*
completely: *pantelôs*
compose: *suntithesthai*
composed of, be: *sunkeisthai*
composite: *sustasis*
composition: *sustasis*
comprehend: *sunidein*

compress: *piloun, puknoun, sumpileisthai*
compression: *pilêsis*
conceive: *epinoein, huponoein*
concept, conception: *ennoia*
conclude: *sunagein*
conclusion: *sumperasma*
condemn: *katapsêphizesthai*
condense: *puknoun*
confusedly: *amudrôs*
conjunction: *sundesmos*
connected with, be: *hupopiptein*
consequence: *sunêmmenon*
 be a consequence: *sumbainein*
consequence of, be a: *akolouthein*
consider: *noein, nomizein*
constituted, be: *sunistasthai*
consumed, be: *ekpurênizein*
contain: *emperiekhein, enapolambanein*
container: *angeion*
continuity: *sunekheia*
continuous: *athroos, sunekhês*
continuously: *sunekhôs*
contract: *aniesthai, diastellein, sunagein, sustellein*
contraction: *sunagôgê*
contrariwise: *antistrophôs, empalin*
contribute: *sumballesthai*
conversely: *antestrammenôs*
co-perception: *sunaisthêsis*
cork: *phellos*
correctly: *kuriôs*
cosmic: *kosmikos*
count: *aparithmein*
cover: *diienai*
craft: *tekhnê*
creation: *genesis*
credibility: *pistis*
creep among: *parempiptein*
criticise: *dielenkhein, elenkhein*
crowd: *plêthos*
cube: *kubos*
cubit: *pêkhuaios*
cup: *potêrion*
cupful: *kuathos, kuathiaios*
curve: *kurtotês*
curved: *kurtos*
cut (v.): *temnein*
cut (n.): *tomê*
cut off: *apotemnein*
cut up: *katatemnein*
cylinder: *kulindros*
cynosaura: *kunosaura*

dampness: *hugrotês*
deal with: *dialambanein*
dear: *philos*
decide: *aphorizein*
declare: *apophainesthai*
decrease in size: *meiôsis*
deduce: *sullogizesthai*
define: *diorizein, horizesthai*
definite, be: *perainein*
definition: *horismos, horos, logos*
Democritus: *Dêmokritos*
demolish: *sunanairein*
demolition: *anairesis*
demonstrate: *apodeiknunai, deiknunai*
dense: *pakhus, puknos*
 become denser: *puknoun*
densification: *puknôsis*
density: *puknôsis*
deny: *anairein*
deprive: *sterein*
desire: *ephiesthai*
desire, object of: *orektos*
destroy: *anairein*
device: *mêkhanê*
devoid of, be: *steresthai*
diameter: *diametros*
differ: *diapherein*
difference: *diaphora, heterotês*
 be a difference: *diaphônein*
 lacking differences: *adiaphoros*
different: *diaphoros, heteros*
differentiation: *diaphora*
difficult: *aporos, duskherês*
dimension: *diastêma*
diminish: *summeioun*
dip: *kathienai*
disbelief: *apistia*
discredit: *elenkhein*
discuss: *dialambanein*
discussion: *logos*

disperse: *kataspan*
disprove: *elenkhein*
distance: *diastasis, diastêma*
distinction: *diakrisis*
distinguish: *diakrinein, diapherein, diastellein*
distinguishing: *diakritikos*
diversion: *apagogê*
divide: *diairein, dielein*
  divide up: *katakermatizein*
  easily divided: *eudiairetos*
  divisible: *diairetos*
division: *diairesis, diorisis*
do: *dran, poiein*
do away with: *anairein, aphairein, elenkhein, sunanairein*
double pipe: *diaulos*
doubled: *diplous*
downwards: *katô*
  tendency to carry downwards: *katôphoros*
dry: *xêros*
dryness: *xêrotês*
dust, cloud of: *koniortos*
dyad: *duas*

earth: *gê*
effective: *drastikos*
element: *stoikheion*
empty: *kenos*
end: *eskhatos*
endure: *hupomenein*
enquire: *episkopein, zêtein*
enquiry: *logos, theôrêma*
entangle: *empodizein*
entanglement: *paremplokê*
enter into: *eisienai*
entirely: *pantelôs*
entry: *hodos, pareisdusis*
equal: *isos*
equal, when day and night are: *isêmerinos*
equalising: *isôsis*
equality: *isotês*
equip: *exartan*
erratic, be: *planasthai*
error: *planê*
escape: *diapheugein, ekpheugein*
evenly: *homalôs*
everywhere: *pantakhou*
  from every angle: *pantakhou*
  in every direction: *pantakhou*
evidence: *tekmêrion*
example: *lemmation, paradeigma*
exceed: *huperballein*
excess: *huperokhê*
exchange: *ameibein, antiparakhôrein, methistanai*
exhaust: *exasthenein*
exhausted, be: *ekluein*
exist: *huparkhein, huphistanai, sunistasthai*
exist in: *enuparkein*
existence: *huparxis*
expansion: *diastolê*
expelled, be: *ekpurênizein*
expend: *katadapanan*
explanation: *exêgêsis*
extend: *epiteinein*
extended: *diastatos*
extension: *diastasis, diastêma*

fall: *sumpiptein, sunizanein*
fall into: *empiptein, peripiptein*
fall out: *ekpiptein*
fall out beyond: *huperekpiptein*
fall under: *hupopiptein*
fall upon: *prospiptein*
false: *pseudos*
falsity: *pseudos*
fanciful: *plasmatôdês*
fast: *oxeôs, takhus*
fill: *apoplêroun, plêroun*
fill up: *anaplêroun, sumplêroun*
filled: *mestos*
filling: *plêrôsis*
filling up: *anaplêrôsis*
final: *teleutaios*
final (cause): *telikos*
find out: *heurizein*
fine: *leptos*
fineness: *leptotês*
finger: *daktulos*
fire: *pur*
fish: *ikhthus*
fisherman: *halieus*

fit: *epharmozein*
fittingly: *epitêdes*
flame: *lignos*
flat: *platus*
flow: *rheuma*
foam: *aphros*
follow: *hepesthai, sumbainein*
following: *akolouthôs*
foot: *pous*
force: *bia, energeia, hormê, rhopê*
with equal force: *isorrhopos*
force out: *exakontizein*
forced: *biaios*
forced, be: *biazesthai*
forcibly: *biai*
form: *tropos*
foundations: *themelion*
fourfold: *tetraplasios*
freezing: *psuxis*
friction: *paratripsis*
fulfil: *plêroun*
full: *plêrês*
fully: *entelôs*
function: *khreia*
furthest: *exôtatos*

gap: *dialeimma*
gently: *hupheimenôs*
genuine: *kuriôs*
get: *lambanein*
get more: *prokoptein*
get outside of: *existasthai*
get rid of: *anairein*
give: *apodidonai, didonai, endidonai, parekhein*
give birth to: *gennan*
give way: *existasthai, hupexienai, hupexistanai, methistanai*
give way to: *hupeikein, metienai*
giving back: *antapodosis*
glass: *speklos*
glide off: *exolisthainein*
go: *hienai, kinein, metabainein*
go into: *khôrein*
go out: *exienai*
go through: *dierkhesthai, diienai*
go with: *hepesthai*
goal: *skopos, telos*
goat-stag: *tragelaphos*
good: *agathos*
gold: *khrusos*
grant: *didonai*
grasp: *ekhein*
growth: *auxêsis*

half: *hemisus*
hand: *kheir*
hang on to: *artan*
happen: *genesthai, sumbainein*
hard: *sklêros*
hardness: *sklêrotês*
harshness: *deinotês*
head: *kephalê*
hearing: *akouê*
heat (n.): *kauma, thermotês*
heat (v.): *thermainein*
heaven: *ouranos*
heavenly: *ouranios*
heavy: *barus*
hide: *kruptein*
hide in: *enkriptein*
hinder: *empodizein*
hold: *enapolambanein*
hold up: *anekhesthai*
hole: *opê*
hot: *thermos*
hour: *hora*
house: *oikia*
hurry: *horman*
hypothesis: *hupothesis*
lay down a hypothesis: *hupotithesthai*
hypothetical: *hupothetikos*

idea: *ennoia, huponoia*
imagination: *phantasia*
imaginative ideas, have: *periphantazesthai*
imagine: *phantazesthai*
immediately: *amesôs*
impetus: *arkhê*
imply: *sunagein*
impossible: *adunatos*
impulse: *rhopê*
include: *enapolambanein, perilambanein,*

sumperilambanein
inclusive: *periektikos*
incorporeal: *asômatos*
increase (v.): *auxanein, prokoptein*
increase (n.): *auxêsis, epiteinein*
indicate: *sêmainein*
inequality: *anisotês*
ingenious: *mêkhanikos*
inseparable: *akhôristos*
insoluble: *aporos*
instantaneously: *akhronôs*
instrumental: *organikos*
interchange (of places): *antiperistasis*
intersperse: *enkataspeirein, paraspan*
interval: *diastêma*
 with interval: *diastêmatikos*
introduce: *eisagein, epagein, suneisagein*
invention: *plasma, plasmatôdês*
irrational: *alogos*
iron: *sidêros*
ironically: *eirôneuomenôs*

jar: *pithos*
join: *proskhôrein*
joining together: *sumplokê*
justifiable: *dikaios*

kind: *tropos*
 of the same kind: *sungenês*
kinetic: *kinêtikos*
kinship: *sungeneia*
knife: *makhairios*

lack: *steresthai*
largeness: *megethos*
laughable: *geloios, katagelastikos*
lead (v.): *apagein*
lead (n.): *molibdos*
leaden weight: *molibdis*
leaf: *petaôdês*
learn: *manthanein*
leave: *kataleipein*
leave out: *paralimpanein*
left: *aristeros*
length: *mêkos*
less than, be: *hupeikein*
let down: *kathienai*
letter: *katagraphos*
lie beside: *parakeisthai*
lie on: *epigenesthai*
lift: *phora*
light: *kouphos*
lightness: *kouphotês*
like: *homoios*
limit (v.): *perainein*
limit (n.): *peras*
limitless: *apeiros*
line: *grammê*
liquid: *hugros*
little: *brakhus*
logical argument: *sullogismos*
look at: *skopein, theôrein*
look into: *episkopein*
look round at: *periathrein*
loss: *apobolê*
ludricous: *geloios*

make: *poiein*
man: *anêr, anthrôpos*
mark: *sêmeion*
mathematical: *mathêmatikos*
matter: *hulê*
mean: *sêmainein*
meaning, give a new to: *kainotomein*
meaning, with the same: *isodunamos*
meanings, with several: *homônumos*
mechanical: *banausos*
meet: *apantan, tunkhanein*
mind: *dianoia*
mingle in: *sunkheein*
miss: *diamartanein*
mix with: *enkerannunai*
moisten: *hugrainein*
monad: *monas*
motion: *phora*
motionless: *akinêtos*
motionless, be: *êremein*
mouth: *stoma*
move: *kinein, metakineisthai*
 easily moved: *eukinêtos*

move along with: *sunkinein*
move through: *khôrein*
movement: *phora*
muddy: *telmatôdês*
mutual exchange: *antiperistasis*

name (n.): *onoma*
name, give a (v.): *kalein*
narrow: *brakhus, stenos*
natural: *emphutos, phusikos, phusei, kata phusin*
naturally: *autophuês, kata phusin*
nature: *phusis*
  be suited by nature: *phunai*
neatly: *kharientôs*
necessary: *anankaios*
necessity: *anankê*
need in addition: *prosdeisthai*
net: *linon*
notch: *gluphis*
note (v.): *sêmainein*
note (n.): *skhôlê*
nourishment: *trophê*
number: *arithmos*

object: *telos*
object of desire: *ephetos*
objection: *apantêsis*
obscure: *adêlos, aphanês*
observe: *horan, sunaisthanesthai*
obvious: *dêlos, enargês, eudêlos*
  very obvious: *prodêlos*
occupy: *ephaptesthai, katalambanein, katekhein*
occur: *sumbainein*
olive (oil): *eliaios*
olive stone: *purên*
one and a third: *epitritos*
openly: *prophanôs*
opinion: *doxa, ennoia*
oppose: *antipaiein, enistanai*
opposed to, be: *antikeisthai*
opposing thrust: *antôsthêsis*
opposite: *antikeimenôs, enantios*
  move in opposite direction: *antikineisthai*
order: *epitagma*
origin: *arkhê*
outermost: *eskhatos*
outside: *perix*
outside, from: *exôthen*
overflow: *saleuein*
overlook: *periidein*
overthrow: *kataballein*
own: *oikeos*
  its own: *idios*

part (v.): *merizein*
part (n.): *meros, morion*
parts, having large: *megalomerês*
parts, with thin/fine: *leptomerês*
parts, with very small: *mikromerês*
parts, with coarse: *pakhumerês*
pass round, pass by: *antimethistasthai*
pass through: *khôrein*
passage: *diodos, parodos, khôrion*
passing away: *phthora*
path: *keleuthos*
peculiarity: *idiotês*
perceptible: *aisthêtos*
perception: *aisthêsis*
periphery: *periphereia*
person: *prosôpos*
persuasive: *pithanos*
persuasive power: *pithanotês*
perversion: *ektropê*
philosopher, natural: *phusikos*
pieces in small: *leptomerôs*
pierce: *tetrainein*
pipe: *surinx*
place (n.): *topos*
place: *apotithesthai*
place in: *entithenai*
plane: *epiphaneia*
plausible: *pithanos*
point: *sêmeion, stigmê*
pointed: *oxus*
pole: *polos*
pore: *poros*
positing: *thesis*
position: *taxis*
posterior: *husteros*
postulate: *hupotithesthai*
pot: *lekanê*
potentiality: *dunamis*

pound, of a: *litriaios*
pounds, of two: *dilitraios*
  half-pound: *hemilitraios*
pour into: *epikheein*
pour out: *ekkheein*
power: *dunamis, rhumê*
power to, give: *apobiazesthai*
practise: *metienai*
precede: *prouparkhein*
precisely: *akribôs, horismenôs*
preferentially: *proêgmenôs*
premise, additional: *proslêpsis*
present, be: *pareinai*
preserve: *diaphulattein*
press: *eirgein, piloun*
press out: *ekthlibein*
pressure: *pilêsis, thlipsis*
prevent: *empodizein, kôluein*
privation: *sterêsis*
probable: *eulogos*
problem: *aporia, aporos, problêma*
proceed: *proagein, proienai*
produce: *apotelein, sunagein*
progress: *epidosis, probainein*
project, be a: *prokeisthai*
proof: *deixis*
proper: *kuriôs, oikeos*
properties, without: *apathês*
proportion: *analogia*
proportionate: *analogos*
proposal: *theôrêma*
proposed, be: *prokeisthai*
proposition: *prokeimenon*
prove: *deiknunai, elenkhein*
provide: *apodidonai, parekhein*
pull: *kataspan*
pursue: *diôkein*
push: *ôthein*
put: *suntassein*
put in: *ballein, emballein,*
    enapotithesthai, entithenai
putting together: *sunkritikos*
Pythagoreans: *Puthagoreioi*

quality: *poiotês*
  have qualities: *poieisthai*
  without quality: *apoios*
quantity: *megethos*
  give quantity: *posoun*
question: *aporein*
quick: *takhus*

ram: *krios*
rare: *manos*
rarefaction: *manôsis*
rarefy: *manoun*
rate, at the same: *isotakhôs*
ratio: *logos*
ray: *aktis*
reach out for: *horegein*
reality: *hupostasis*
reason: *aitia, logos*
reasonable: *eulogos*
reasoning: *logos*
receive: *dekhesthai, diadekhesthai*
receptive of: *dektikos*
record: *hupomimnêskein*
recount: *aparithmein*
reduce: *katadapanan*
redundant: *mataios, perittos*
refute: *elenkhein*
relate to: *epharmozein*
related to, be: *pheresthai*
relation: *analogia, skhesis*
relationship: *sungeneia*
remain: *menein*
remark: *rhêsis*
repetition: *epanalêpsis*
reply: *apantêsis*
reputation: *hupeilêmmenon*
resist: *empodizein*
resistance: *antitupia*
resistant: *antitupos, empodistikos*
resolve: *diakrinein*
rest: *stasis*
result: *sumbainein*
revolution: *dinê*
ride on: *epokheisthai*
right: *dexios*
river: *potamos*
roll, difficult to: *duskuliskos*
roll, easy to: *eukulistos*
roof: *orophê*
room: *khôra*
  have room for: *khôrein,*
    metakhôrein

run: *trokhein*
run back: *palindromein*
rush: *hormê*

safeguard: *phulassein*
sailor: *nauta*
save: *sôizein*
scatter: *diaskedannunai, enkataspeirein, kataspan, skedannunai*
seal (v.): *apophrassein*
seal (n.): *kêtos*
search: *zêtêsis*
section: *logos*
see: *horan*
  what sees: *opsis*
see as: *theôrein*
seen: *horatos*
seize: *katalambanein*
self-evident: *enargês*
semicycle: *hêmikuklion*
sense: *aisthêsis, tropos*
  in a wide sense: *koinos*
separate (adj.): *khôristos*
separate (v.): *apokrinein, khôrizein*
separating, be: *diistasthai*
separation: *khôrismos*
set out: *ektithenai, prolambanein, protithenai*
set up: *tithesthai*
shape: *skhêma*
  of similar shape: *homoioskhêmos*
  with equal shapes: *isoskhêmôn*
share: *moira*
sharpness: *oxutês*
shoot: *toxeuein*
shoot out: *exakontizein*
show: *apodeiknunai, deiknunai*
show as well: *sunapodeiknunai*
shrink: *sunizanein*
shrinkage: *sunizêsis*
side: *plagios*
side by side: *parallêlos*
similar: *sungenês*
similar, be: *exomoiousthai*
similarity: *homoiotês*
simple: *haplous*
simply: *haplôs*
size: *megethos*
skin: *askos*
slacken: *khalan*
slow: *aphauros, bradus*
slowness: *bradutês*
smaller, make: *meioun*
smallness: *smikrotês*
smoke: *kapnos*
  like smoke: *atmidôdês*
snatcher: *harpagion*
soft: *malakos*
softness: *malakotês*
solid: *sôma, stereos, sterros*
solution: *diakrisis, lusis*
solve: *epiluein, luein*
sophist: *sophistês*
sound: *phônê*
sound: *psophos*
sounding: *psophêtikos*
space: *khôra*
space, of: *topikos*
speak on behalf of: *apologeisthai*
speed: *phora, takhos*
spend: *dapanasthai*
sphere: *sphaira*
  like a sphere: *sphairoeidês*
spherical: *sphairikos*
split: *skhizesthai*
sponge: *spongos*
spot: *topos*
spread out: *diastastos*
spring: *ear*
spurious, treat as: *notheuein*
squeeze: *piloun, thlibein*
stade long, a: *stadiaios*
stand: *histanai*
start (n.): *arkhê*
start (v.): *arkhesthai*
state: *hexis, sustasis*
statement: *apophansis*
stay the same: *menein*
steam: *atmis*
  turn into steam: *exatmizein*
stir: *kiknein*
stomach: *gastêr*
stone: *lithos*
  of stone: *lithinos*
stone (olive): *gigarton*

stop: *histanai, kôluein, phrassein*
stopping: *stasis*
story: *logos*
straight: *euthus*
straightway: *autothen*
strength, of equal: *isosthenês*
strict sense, in the: *kuriôs*
strike: *prosballein*
strike on: *prospiptein*
string: *neura*
strip: *gumnazein*
stuff: *ousia*
stupid: *anoêtos*
subject: *theôrêma*
substance: *ousia, sôma*
substantial: *iskhuros*
substrate: *hupokeimenon*
sufficient, be: *exarkein*
suggest: *hupotithesthai, tithesthai*
suggestion: *hupothesis*
sui generis: *anomoiogenês*
summer: *theros*
sun: *hêlios*
  of the sun: *hêliakos*
support: *marturein, sunagoreuein*
suppose: *hupolambanein, huponoein, hupotithesthai, nomizein*
supposition: *hupolêpsis*
surface: *epipedon, epiphaneia*
surpass: *huperekhein*
surround: *periballein, periekhein*
swell: *kumainein*
swim: *nôkhesthai*
swimming pool: *kolumbêthra*
symbol: *sumbolon*
symbolic way, in a: *sumbolikôs*

take: *katalambanein, lambanein*
take away: *huphairein*
take hold of: *ephaptesthai*
take in: *enapolambanein*
take in in addition: *proslambanein*
take up: *analambanein*
take with it: *summetapherein*
talk about: *dialegesthai*
tangible: *haptos*
tangle: *sumpodizein*
teach: *didaskein*
tear apart: *diaspeirein*
tear away: *apospan*
ten thousand: *murios*
tenfold: *dekaplasios*
tension: *tonos*
test: *elenkhein*
testify: *marturein*
text: *graphê, lexis*
then: *akolouthôs*
theory: *doxa, hupothesis*
thickness: *pakhutês*
thing: *pragma*
think: *axioun, doxazein*
think about: *epinoein*
think of: *noein, tithesthai*
thinner, to become: *leptunein*
thinner, to make: *manoun*
thought: *dianoia, ennoia, epinoia, logos*
three directions, in: *trikhêi*
three ways: *trikhêi*
throw: *rhiptein*
throw off: *apoballein*
thrust (n.): *ôsis, ôthêsis*
thrust (v.): *ôthein*
thrust back: *anakrouein*
time: *hora, kairos, khronos*
  taking more time: *khronios*
tip: *akros*
topical: *topikos*
touch (v.): *ephaptesthai, haptesthai*
touch (n.): *haphê*
track: *sêmeion*
transfer: *metaballein*
trap: *aporos*
traverse: *diexienai*
trend: *phora*
triad: *trias*
triple pipe: *triaulos*
true: *alêthês*
try: *epikheirein*
turn (v.): *antistrephein, strephein, trepesthai*
turn (n.): *tropê*
twist: *streblein*
twofold: *diplasios, dittos*
two-thirds: *dimoiros*

unbelievable: *apithanos*
unclearly: *asaphôs*
undergo: *hupomenein*
underlie: *hupokeisthai*
understand: *epistasthai, exakouein, sunidein*
unequal: *anisos*
unequal speed: *anisotakhos*
union: *sustasis*
unit: *monas*
universe: *kosmos*
  of the universe: *kosmikos*
unknown: *agnooumenos*
unnatural: *para phusin*
unprovable: *anapodeiktos*
unspecified: *aprosdioristos*
unstop: *apophrassein*
upwards: *anô*
  tending to go upwards: *anôphoros*
urge: *hormê*
use (n.): *khreia*
use (v.): *khrêsthai*
use in addition to: *proskhrêsthai*
useless: *mataios, perittos*

vain: *mataios*
vapour (n.): *pneuma*
  turn into vapour (v.): *empneumatoun*
  turning into vapour (n.): *empneumatôsis*
vessel: *angeion, angos*
view: *doxa*
void: *kenos*
volume: *onkos*
  with equal volume: *isonkos*

walk: *badizein*
wall: *toikhos*
want: *boulesthai*
water: *hudôr*
way: *tropos*
way out: *diexodos*
weak: *asthenês*
weave: *prosplekein*
weight: *baros, barutês*
well-known: *endoxos*
wet: *hugros*
wheel in: *pareiskuklein*
whirl around: *peristrephein*
white: *leukos*
whiteness: *leukotês*
whole: *holos*
wholeness: *holotês*
wide: *platus*
widthless: *aplatos*
wine: *oinos*
wine (new): *gleukos*
winter: *kheimôn*
within, from: *oikothen*
without: *dikha, erêmos*
witness (n.): *martus*
  bear witness (v.): *marturein*
wonder: *aporein*
wood: *xulos*
wooden: *xulinos*
wool: *erion*
word: *logos, onoma, phonê, rhêtos*
words: *lexis*
work: *pragmateia*
work out: *noein*
writings: *logos*
wrongly: *katakhrêstikôs*

yawning gulf: *akhanês*
yield: *eikein, pareikein, peithesthai;* (to one another) *antiparakhôrein*
yielding: *hupeiktikos*

zodiac: *zôidion*

Greek-English Index

Index of Passages

Subject Index

Made in the USA
Monee, IL
22 October 2025

32655126R00083